WILD BERRIES
of BRITISH
COLUMBIA

Fiona Hamersley Chambers

Lone Pine Publishing

The Publisher: Lone Pine Publishing
10145 – 81 Avenue
Edmonton, Alberta T6E 1W9
Website: www.lonepinepublishing.com

Library and Archives Canada Cataloguing in Publication

Hamersley Chambers, Fiona, 1970–
    Wild berries of British Columbia / Fiona Hamersley Chambers.

Includes index.
ISBN 978-1-55105-865-8

    1. Berries—British Columbia—Identification. I. Title.

QK203.B7C527 2011   581.4'6409711   C2010-907241-3

Editorial Director: Nancy Foulds
Project Editor: Gary Whyte
Editorial Support: Kelsey Everton, Nicholle Carriere
Photo/Illustration Coordinators: Kelsey Everton, Genevieve Boyer
Photo Cataloguing: Randy Kennedy
Photo Correction: Gary Whyte, Kamila Kwiatkowska
Production Manager: Gene Longson
Book Design: Lisa Morley, Gerry Dotto
Layout & Production: Lisa Morley, Janina Kuerschner
Cover Design: Gerry Dotto
Author Photo: Vicki Pallan Photography

Photographs and illustrations in this book are used with the generous permission of their copyright holders. Illustration and photo credits are located on p. 191, which constitutes an extension of this copyright page.

DISCLAIMER: This guide is not meant to be a "how-to" reference guide for consuming wild berries. We do not recommend experimentation by readers, and we caution that many of the plants in Canada, including some berries, are poisonous and harmful. The authors and publisher are not responsible for the actions of the reader.

We acknowledge the financial support of the Government of Canada through the Canada Book Fund (CBF) for our publishing activities.

PC: 19

Dedication

To three incredible women who I am fortunate to have in my life: my mother Sarah Richardson for being so wonderfully enthusiastic about plants and for doing lots of quality child care while I write, my second mum Vicky Husband for showing me what you can do if you don't take "no" for an answer and persist against the odds, and to Nancy Turner for showing me what you can accomplish by always being positive and kind. And most of all to my boys Hayden and Ben. You guys are the best foragers and berry testers I know and you give me such joy.

Acknowledgements

The following people are thanked for their valued contribution to this book:

- ❧ The many photographers who allowed us to use their photographs

- ❧ Lone Pine's editorial and production staff;

- ❧ and most importantly, the native peoples, settlers, botanists and writers who kept written records or oral accounts of the many uses of the berries featured in this book. I am particularly indebted to my family, friends and teachers who have taught me so much about our native plants and their uses.

Fiona Hamersley Chambers
School of Environmental Studies, University of Victoria
Pacific Rim College
www.metchosinfarm.ca

CONTENTS

THE BERRIES

List of Recipes

Plants at a Glance

TREES AND SHRUBS

Pacific Crab Apple p. 28

Junipers p. 34

Hawthorns p. 40

Arbutus p. 32

Mountain Ash p. 44

Wild Roses p. 46

Red Cherries p. 52

Chokecherry p. 56

Indian-plum p. 60

Smooth Sumac p. 62

Cloudberry p. 72

Thimbleberry p. 74

Blackberries p. 64

Raspberries p. 68

Salmonberry p. 76

Oregon-grapes p. 78

Currants p. 82

Gooseberries p. 88

Prickly Currants p. 94

6

 Saskatoon p. 98

 Dogwoods p. 100

 Huckleberries p. 104

 Blueberries p. 112

 Cranberries p. 118

 False-wintergreens p. 122

 Bearberries p. 126

 Salal p. 130

 Black Crowberry p. 132

 Elderberries p. 134

Bush Cranberries p. 138 Soapberry p. 142

Silverberry p. 144

FLOWERING PLANTS

 One-flowered Clintonia p. 146

 Twisted-stalks p. 148

7

False Solomon's-seals p. 156

Fairybells p. 158

Lilies-of-the-valley p. 152 Strawberries p. 160 Bunchberry p. 164

Poisonous Plants

English Holly p. 166 Cascara p. 168 Pacific Yew p. 170

Devil's Club p. 174

Honeysuckles p. 178

Poison-ivy & Poison-oak p. 172 Common Snowberry p. 180 Red Baneberry p. 182

British Columbia (with cities and mountain ranges)

ST. ELIAS MTNS.

CASSIAR MTNS.

ROCKY MTNS.

Cassiar   Coal River

Fort Nelson

COAST MTNS.

Telegraph Creek

SKEENA MTNS.

OMINECA MTNS.

Dawson Creek

Smithers

Prince Rupert   Fort St. James

Prince George

QUEEN CHARLOTTE MTNS.

ROCKY MTNS.

Valemount

Bella Coola   Quesnel

COAST MTNS.

100 Mile House   Golden

Kamloops   Revelstoke

Kelowna

Vancouver

Victoria

Creston

CASCADE MTNS.

VANCOUVER ISLAND RANGES

COLUMBIA MTNS.

Introduction

Woodland strawberry (*Fragaria vesca*)

It's difficult to find someone who does not enjoy eating a berry. Juicy, sweet, tart, sometimes sour, generally bursting with flavour and very good for you—wild berries are gifts from the land, treasures to be discovered on a casual hike or potentially a lifesaving food if you're unfortunate enough to get lost in the woods. Berries have a long history of human use and enjoyment as food and medicine, in ceremonies and for ornamental and wildlife value. Our ancestors needed to know as a matter of survival which berries were edible or poisonous, where they grew and in which seasons and how to preserve them

for non-seasonal use. These early peoples often went to great lengths to manage their wild berry resources: pruning, coppicing, burning, transplanting and even selectively breeding some wild species into the domesticated ancestors of many of our modern fruit varieties.

Today, many of us live in urban environments where the food on our plate and in our pantries comes from great distances away. The first strawberry is no longer an eagerly awaited and delectable harbinger of the summer to come. Rather than a fleetingly sweet June moment, these fruits are now available on our grocery store shelves almost year-round.

A sad result of this convenience and lack of seasonality is that this store-bought fruit has little resemblance to its forebears. Grocery store strawberries, for example, are generally not properly ripe, don't taste of much and are not loaded with nutrients. As we become more and more disconnected from our food sources, it is even possible that we are forgetting what a "real" berry tastes like. Perhaps part of the exceptional taste of wild fruit is the thrill of the hunt and the discovery of a gleaming berry treasure hanging—sometimes in great profusion—from a vine or bush. These wild berries are only available for a short time during the year and we must increasingly travel to find them growing in their native state. We must make an effort to discover them in the wild or find a reputable source for those plants that will grow in our home gardens.

The wild berries described in this book are, for the most part, not available in stores. When they are, they are often very expensive. A berry in its prime state of ripeness is juicy and delicate, and therefore does not travel well. What a pity, as slowly savouring one of these fruits at its peak of perfection plucked fresh off the plant is one of the great joys in life. What better way to spend a warm summer's day than wandering hillsides, country roads, or forest edges with friends and family in search of these delectable morsels? Wild berry gathering builds community and family and is a great way to connect you and your children to nature. In winter, a spoonful of these frozen or preserved wild fruits will bring back the taste of summer for a delicious moment. My young boys love to go berry gathering with me and are proud when they share the jams and baking that they have made. It is my hope that, whether you are a seasoned gatherer or a new enthusiast, this book will help guide you to experience and share in this wonderful and generous gift of nature.

Why Learn to Identify and Gather Wild Berries?

Berries gathered in the wild generally have superb flavour and can be gathered when they are properly ripe. These fruits are not only delicious, but contain important nutrients and phytochemicals (anti-cancer compounds) that are increasingly lacking from our commercially available fruits. Many wild berries are high in vitamin C and also contain trace elements, carbohydrates, proteins, and important nutrients such as iron, calcium, thiamine and vitamin A. While most people will obtain this guidebook in order to enjoy wild berries on hikes and outings, the information that you learn here could

Wild red raspberry (*Rubus idaeus*)

also save your life if you ever get stranded or lost in the backcountry. Be warned, though! Gathering wild berries can also be considered a dangerous "gateway" into the more complex realm of preserving and cooking with these fruits as well as growing them in your own backyard. Once you start on this journey it can become rather addictive and even spread to friends and family!

What is Not Covered in this Guide?

Although this book should enable you to identify most native berry species in your region, it is not intended as a complete reference guide. A section on references and further reading is provided for those wishing to study these plants in greater detail. Some berry species are so rarely found or have such a restricted range that it would not be useful to include them here. There are many excellent resources already available to help you understand the cultivation and use of domesticated fruit species, so these are also not covered. Nuts, seeds and cones are not considered "berries" in the common sense of the word, so these are excluded.

What is a "Berry"?

In this guide, "berry" is used in the popular sense of the word, rather than in strictly botanical terms, and includes any small fleshy fruit. Technically, a "berry" is a fleshy, simple fruit produced from a single ovary that contains one or more ovule-bearing structures (carpels) that each contains one or more seeds. The outside covering

(the endocarp) of a berry is generally soft, moist and fleshy, most often in a globular shape. Roughly translated, a berry is really a seed(s) packaged in a tasty moist pulp that encourages animals to eat the fruit and distribute the seeds far and wide from the parent plant so that these offspring can grow and flourish. "True berries" include currants, huckleberries, blueberries and grapes.

Botanically, however, what we call a "berry" often includes simple fleshy fruits such as drupes and pomes. The botanical definitions of different types of fruit are provided below for general interest and are also sometimes mentioned, where appropriate, in the text.

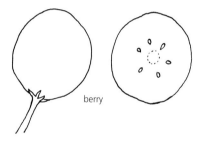

berry

A "drupe" is a fleshy stone fruit that closely resembles a berry but has a single seed or stone with a hard inner ovary wall that is surrounded by a fleshy tissue. Wild fruit in this category include high bush cranberries and bunchberries; some domestic fruit examples are cherries or plums.

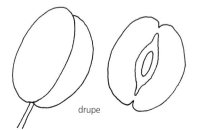

drupe

Himalayan blackberry (*Rubus armeniacus*)

A "compound drupe" or "aggregate" fruit ripens from a flower that has multiple pistils, all of which ripen together into a mass of multiple fruits, called "drupelets." A drupelet is a collection of tiny fruit that forms within the same flower from individual ovaries. As a result, these fruit are often crunchy and seedy. Wild examples include raspberries and blackberries. Cultivated examples include tayberries, loganberries and boysenberries.

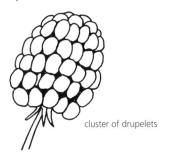

cluster of drupelets

A "multiple fruit" is similar to an aggregate, but differs in that it ripens from a number of separate flowers that grow closely together, each with its own pistil (as opposed to from a single flower with many pistils). Mulberry is the only native Canadian example of a multiple fruit. Tropical examples include pineapples and figs.

An "accessory fruit" is a simple fruit with some of its flesh deriving from a part other than the ripened ovary. In other words, a source other than the ovary generates the edible part of the fruit. Other names for this type of fruit include pseudocarp, false fruit or spurious fruit. A "pome" is a sort of accessory fruit because it has a fleshy outer layer surrounding a core of seeds enclosed with bony or cartilage-like membranes (it is this inner core that is considered the "true" fruit). Serviceberries and hawthorns are wild examples of pomes; apples and pears are domestic examples. Another type of accessory fruit is the strawberry; the main part of the fruit derives from the receptacle (the fleshy part that stays on the plant when you pick a raspberry) rather than from the ovary. Wintergreen is also an example of an accessory fruit type as the fruit is really a dry capsule surrounded by a fleshy calyx.

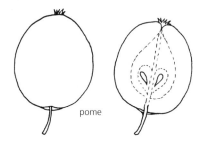

pome

A "cone" is a fruit that is made up of scales (sporophylls) that are arranged in a spiral or overlapping pattern around a central core, and in which the seeds develop between the scales. The juniper is an example of a species that produces cones.

cone

A "hip" has a collection of bony seeds (achenes), each of which comes from a single pistil, covered by a fleshy receptacle that is contracted at the mouth. The rose hip (which is also an accessory fruit) is our only example of a hip.

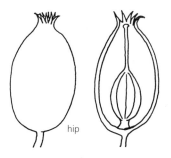

hip

The Species Accounts

In this book, species are organized by growth form into three main sections—Trees and Shrubs, Flowering Plants, and Poisonous Plants. Closely related or similar plants are grouped together for comparison and the section on poisonous plants is conveniently located at the end of the book.

This book includes plants common in British Columbia that have been used by people, both in ancient times and in the present. Each account has a detailed description for each plant including plant form, leaf structure, habitat and range, and fruit form, colour and season. This description, in addition to colour photographs and illustrations, will help you ensure safe plant and berry identification. Information on traditional and contemporary uses for food, medicines

English hawthorn (*Crataegus monogyna*)

Bunchberry (*Cornus canadensis*)

and material culture are also included for general interest.

The information in each species account is presented in an easy to follow format. In addition to the opening general discussion, each account includes subheadings of Edibility (see below for edibility scale), Fruit (a description of the look and taste of the fruit), Season (flowering and fruiting seasons) and Description (a detailed description of the plant, flowers, habitat and range).

Many species accounts focus on a single species, but if several similar species have been used in the same ways, two or more species may be described together. In these "group" accounts, you will find a general description for the group, but there will also be separate paragraphs in which the individual species in the group are described in specific detail. So, for example, the Hawthorns (*Crataegus* spp.) account describes historical and modern uses in the

opening discussion, then has subheadings of Edibility, Fruit, Season, and Description for all hawthorns. These subheadings are followed by separate paragraphs with important specific information (including identification and location details) for each of the numerous hawthorns in the province (black hawthorn, English hawthorn, red hawthorn, Suksdorf's hawthorn).

Where appropriate, you will find an "Also called" subheading that describes other common and scientific names for each species. These "Also called" names are found below the main account title and at the end of individual species description paragraphs inside the account.

Edibility Scale

All accounts contain a useful scale of edibility for each species. Although we have a wonderful variety of native berries, it is useful to know which ones are worth our time pursuing; which ones, although considered "edible," are better left for the birds or as a famine food; and which ones are toxic or poisonous.

Highly edible describes those berries that are most delicious and are well worth gathering and consuming. A wild strawberry or serviceberry is considered highly edible.

Edible describes those berries that are still tasty, but not as good as the prime edible species. An example of such a fruit is bunchberry.

Not palatable describes berries you can eat without any ill effects but are perhaps not worth the effort to harvest given their lack of flavour, their bitterness, relatively large seeds or lack of fleshiness. It is useful to know about these species in case you are desperate to snack on something in the woods, but they are not berries that you would actively

Devil's club (*Oplopanax horridus*)

gather to make a pie! Silverberry is an example of such a species.

Edible with caution (toxic) are berries that are palatable, but have differing reports as to their edibility, or perhaps they are only toxic if you eat large amounts or if they haven't been prepared properly or are unripe. Berries of our native juniper species are an example under this category.

Poisonous berries are ones that are definitely poisonous and should not be eaten. An example of a poisonous berry is devil's club.

Season

The season given for flowering and fruit production for each species is an average. Specific microclimates like deep valley bottoms or high mountaintops will necessarily produce a wide range of flowering and fruiting variability for the same plant. Berry plants also produce fruit of differing quality and quantity from year to year, depending on factors such as plant age and health, changes in temperature and moisture, or insect infestation. Some berries, such as rosehips, are best harvested later in the year after the first frost sweetens the fruit.

Description

The plant description and the accompanying photos and illustrations are important parts of each species account. Each plant description begins with a general outline of the form of the species or genus named at the top of the page. Detailed information about diagnostic features of the leaves, flowers

Prairie rose (*Rosa woodsii*)

and fruits is then provided. Flowering time is often included as part of the flower description to give some idea of when to look for blooms and a general fruiting season is also included. If two or more species of the same genus have been used for similar purposes, several of the most common species may be illustrated and their distinguishing features described.

Plant characteristics such as size, shape, fruiting, colour and hairiness vary with season and habitat and with the genetic variability of each species. Identification can be especially tricky when plants have not yet flowered or fruited. If you are familiar with a species and know its leaves or roots at a glance, you may be

able to identify it at any time of year (from very young shoots to the dried remains of last year's plants), but sometimes a positive ID is just not possible.

General habitat information is provided for each species to give you some idea of where to look for a plant. The habitat description provides information about general habitat (e.g., in moist, mossy forest), elevation (e.g., low to montane elevations) and range (e.g., from the northern part of a province to its southern regions). The species ranges and habitats described in this guide were obtained from *Flora of North America*, the United States Department of Agriculture Plants Database website, regional field guides, personal experience and interviews, academic papers and other sources. Despite all due diligence being taken, however, this description is not universal or foolproof. Plants sometimes either grow outside of their reported ranges, or cannot be found within the described habitat. The habitat information included for each species is meant as a general guide only; plants often grow in a variety of habitats over a broad geographical range.

The origin of non-North American species is also noted. The flora of many areas has changed dramatically over the past 200 years, especially in and around human settlements. European settlers brought many plants with them, either accidentally (in ship ballast, packing and livestock bedding) or purposely (for food, medicine, fibre, ornamental value, etc.). Some of these introduced species produce fruit and have thrived, and some are now considered weeds on disturbed sites across much of Canada. An example of such an introduced species is European mountain ash.

European mountain ash (*Sorbus aucuparia*)

What's In a Name?

Both common and scientific names are included for each plant. Scientific names are from *Flora of North America*, for families already completed (*Flora of North America* is a work in progress). For other plant families, scientific names follow *Flora of Canada* (Scoggan 1978–1979). Common names are also largely from *Flora of North America* and *Flora of Canada*.

Common names are often confusing. Sometimes, the same common name can refer to a number of different, even unrelated, species. And, at the same time, one common name can even refer to a plant that is edible and to a completely different and unrelated species that is poisonous! For this reason, the scientific name is included for each plant entry.

The two-part scientific name used by scientists to identify individual plants may look confusing, but it is a simple and universal system that is worth taking a few moments to learn about. Swedish botanist Carolus Linnaeus, who lived from 1707 to 1778, first suggested a system for grouping organisms into hierarchical categories and it is still essentially the same today, almost 300 years after he first developed it! His system differed from other contemporary ones in that it used an organism's morphology (its form and structure) to categorize a species, with a particular emphasis on the reproductive parts, which we now know are the most ancient part of any plant. Another significant benefit of this hierarchical system is that it groups plants into families so that we can better under-stand and see how they are related to each other. For example, both oval-leaved blueberry (*Vaccinium ovalifolium*) and black huckleberry (*Vaccinium membranaceum*) are related cousins in the heath (Ericaceae) family. In another example, Linnaeus' system shows us that roses are botanically related to apples—both are in the Rosaceae family. Since the names of organisms in Linnaeus' system follow a standard format and are in Latin (or a Latinized name formed from other words), they are the same in every language around the world, making this a truly universal classification and naming protocol.

In Linnaeus' system, the species name (the "scientific name") has two parts: (1) the genus, and; (2) a species identifier (or specific epithet), which is often a descriptive word. The first part of the scientific name, the genus, groups species together that have common characteristics. The genus name is always capitalized and both parts of the scientific name are either written in *italics* or <u>underlined</u>. The second part, the specific epithet, which is not capitalized, often describes the physical or other characteristic of the organism, honours a person,

or suggests something about the geographic range of the species. For example, in the scientific name for bunchberry, *Cornus canadensis*, the specific epithet roughly translates as "from Canada." This apt name describes a species that has a wide distribution across our entire country.

It is important to note, however, that botanists do not always agree on how some plants fit into this system. As a result, scientific names can change over time, or there can sometimes be more than one accepted scientific name for a plant. While this is somewhat annoying and may seem redundant, the important thing to remember is that one scientific name will **never** refer to more than one plant. Thus, if you have identified a wild berry as edible and know its scientific name(s), you can confirm that it is indeed edible and not have to worry that this name may refer to another (possibly deadly poisonous!) plant.

Botanists also sometimes further split species into subsets known as varieties. For example, peaches and nectarines are two slightly different varieties of the

Black huckleberry (*Vaccinium membranaceum*)

Prickly currant (*Ribes lacustre*)

peach tree, *Prunus persica*. If you purchase a Harken peach tree at your local plant nursery, the tag should read "*Prunus persica*, variety Harken."

Giving Back to the Plants

While many of our native wild berries grow in profusion, others are threatened by habitat destruction, overharvesting or climate change. In some areas, such as national parks, harvesting is prohibited. Please do not dig up plants from the wild. Most berry species propagate easily from seed or cuttings, and you can also purchase healthy and responsibly produced plants from reputable nurseries. When you harvest native berries in the wild, it's nice to say "thank you" to the plant by weeding back competing species around its base, spreading some of its seeds in similar habitat a short distance from the parent plant, or appropriately pruning the plant if you know the right technique. There is a long history of humans looking after the plants that support us; taking a few moments to continue this tradition and to teach it to our children is time well spent. By learning about our native berry species and harvesting them, we get to know and respect these plants and may even be moved to help protect and propagate them.

A Few Gathering Tips

1. Gather only species that are common and abundant, and never take all the fruit off one plant. Even then, a cautious personal quota will still deplete the plants if too many people gather them in one area. Remember, plants growing in harsh environments (e.g., northern areas, alpine, desert) might not have enough energy to produce flowers and fruits every year. Also, don't forget the local wildlife. Survival of many animals can depend on access to the fruits that you are harvesting.

2. Never gather from plants that grow in protected and/or heavily used areas such as parks and nature preserves. Doing so is not only wrong, but is also often illegal. Be sure to check the regulations for the area you are visiting.

3. Take only what you need, and damage the plant as little as possible. If you want to grow a plant in your garden, try propagating it from seed or a small cutting rather than transplanting it from the wild.

4. Don't take more than you will use. If you are gathering a plant for food, taste a sample. You may not like the taste of the berries, or the fruit at this site may not be as sweet and juicy as the ones you gathered last year.

5. Gather berries only when you are certain of their identity. Many irritating and poisonous plants grow wild in Canada, and some of them resemble edible or medicinal species. If you are not positive that you have the right plant, don't use it. It is better to eat nothing at all than to be poisoned!

Recipes

Simple recipes for cooking, preserving or enjoying berries fresh off the plant are included throughout the book. Every berry gatherer or cook can produce delicious results to enjoy with friends and family in the heat of summer and later during the long winter months. The recipes call for specific berries, but you can experiment by substituting other fruit. For example, you could try replacing blueberries with serviceberries, cranberries or huckleberries.

Dried Fruit

It's hard to beat the flavour of home-dried wild berries. Enjoy these special treats out of the bag or add them to your favourite recipes in place of the usual commercial raisins, dried cranberries or blueberries.

A note on berries that dry well: huckleberries, strawberries, thimbleberries (which are a bit crunchy but have a fabulous flavour), blueberries, saskatoons, cranberries and currants. Berries that do not dry as well: seedier fruit such as blackberries or very juicy fruit such as salmonberries; it is better to mash these types of berries either alone or combined with other fruit and make them into fruit leather. Some fruit, such as elderberries, should be cooked before drying to neutralize the toxins present in the fresh fruit.

If some of the berries are much larger than others, cut them in half. All the berries on a tray should be roughly the same size to ensure even drying. Carefully pick through the fruit to remove insects and debris. Do not wash the berries—it will cause them to go mushy. Lightly grease a rimmed baking sheet and spread the berries on the sheet so that they do not touch each other. Place in a food dehydrator or dry in an oven at 140° F overnight, leaving the oven door ajar to allow moisture to escape. Cool and store in an airtight container or Ziploc® bag.

Frozen Wild Fruit the Easy Way

Freezing is the quickest and easiest way to preserve wild berries, making a wonderful snack any time of the year. Choose the best and ripest fruit and carefully remove all unwanted debris and insects. Some fruit, such as elderberries, should be cooked first to neutralize any toxins. Give dusty berries a quick rinse, though the extra water and handling may bruise the fruits and stick them together during freezing.

Most instructions tell you to freeze berries individually on rimmed baking sheets before packing them in Ziploc® bags. However, I have frozen berries very successfully for years in used milk cartons. Open the carton (1 or 2 L size) fully and wash well in warm, soapy water. Allow to air dry. Cartons with the plastic lid and spout do not work for freezing. I collect cartons during the year, wash them, then store them away for when I need lots of them in the summer. Unless a fruit is particularly mushy (like a very ripe wild raspberry), I simply pick through the fruit to clean it, then gently pour the berries into the carton being careful not to let them pack too hard or crush. Push the top of the carton back together the way it was before opening, then firmly push the top edge so that it folds over flat and indents slightly so that it stays shut. Presto, a sealed container that will never get freezer burn, that is easy to label on the top with a marker pen, and that stacks beautifully in the deep freeze!

To get the frozen fruit out, gently squeeze the carton sides to separate the berries, making it easy to pour out the desired quantity before resealing the carton and returning it to the freezer. If the berries are more firmly attached, simply place the carton on the floor and gently stand on it, turning the sides a few times. As a last resort, peel the carton down to the desired level, cut off the exposed fruit chunk with a sharp knife, and put the remainder of the carton in a Ziploc® bag before replacing it in the freezer. I've successfully re-used the same cartons for many years as long as the fruit inside was not too mushy or difficult to extract.

Northern gooseberry (*Ribes oxyacanthoides*)

A Cautionary Note

If you cannot correctly identify a plant, you should not use it. Identification is more critical with some plants than with others. For example, most people recognize strawberries and raspberries, and all of the species in these two groups are edible, though not all are equally palatable. Plants belonging to the deadly nightshade (solanum) family, however, may be more difficult to distinguish from each other and can range in edibility from highly edible to poisonous. Even the most experienced harvester can sometimes make mistakes. It is important to be certain of a species' identification and any special treatment required before eating a wild berry. Serviceberries, for example, are best cooked to neutralize the poisonous cyanide compounds found in their seeds and many types of under-ripe berries can cause digestive upset or even be poisonous. Some rare

individuals have an allergic reaction to certain berry species.

It is also important to know which parts of the berry are edible. For example, while the fleshy "berry" (it is really an "aril") of the Pacific yew tree is considered edible, eating this fruit is not recommended as the small hard seed contained

Pacific yew (*Taxus brevifolia*)

Common juniper (*Juniperus communis*)

inside is so deadly poisonous that ingesting even a few can cause death! As a general rule, most of our native berry species taste good and are edible. Those berries that have a bitter, astringent or unpalatable taste are telling us that they are toxic or poisonous and that we should not be eating them. These species tend to rely on birds, rather than humans, to eat the fruit and distribute the seeds. The exceptions to these guidelines are the many introduced ornamental plants in our gardens and municipal plantings, some of which have naturalized into the wild and have sweet-tasting fruit. It is not recommended that you sample these non-native fruits without a positive identification. Examples of common poisonous berries are the European lily-of-the-valley and any ornamental yew species.

Finally, some people believe that it is OK to eat berries that they see birds and wildlife enjoying. That is simply not the case so don't test this flawed bit of folklore! Likewise, the fact that a plant has edible fruit does not mean that the plant itself is edible.

Pay attention to where you are harvesting. Fruit growing along the edge of a busy highway or near an industrial area could be contaminated with heavy metals or other pollutants. Municipal plantings might look delicious, but they may be sprayed with pesticides and you might not be welcome to harvest the fruit if it has an ornamental value. Please also remember to harvest on public, not private, lands unless you have received permission from the property owner.

Many plants have developed very effective protective mechanisms. Thorns and stinging hairs discourage animals from touching, let alone eating, many plants. Bitter, often irritating and poisonous compounds in leaves and roots repel grazing animals. Many protective devices are dangerous to humans. The "Warning" boxes throughout the book include notes of potential hazards associated with the plant(s) described. Hazards can range from deadly poisons to spines with irritating compounds in them. These "Warning" boxes may also describe poisonous plants that could be confused with the species being discussed in the account.

The fine line between delicious and dangerous is not always clearly defined. Many of the plants that we eat every day contain toxins and almost any food is toxic if you eat too much of it. Personal sensitivities can also be important. People with allergies may die from eating common foods (e.g., peanuts) that are harmless to most of the population. Most wild plants are not widely used today, so their effects on a broad spectrum of society remain unknown.

As with many aspects of life, the best approach is "moderation in all things." Sample wisely—when trying something for the first time, take only a small amount to see how you like it and how your body reacts.

No Two Plants Are the Same

Wild plants are highly variable. No two individuals of a species are identical and many characteristics can vary. Some of the more easily observed characteristics include the colour, shape and size of stems, leaves, flowers and fruits. Other less obvious features, such as sweetness, toughness, juiciness and concentrations of toxins or drugs, also vary from one plant to the next.

Bitter cherry (*Prunus emarginata*)

Many factors control plant characteristics. Time is one of the most obvious. All plants change as they grow and age. Usually, young leaves are the most tender, and mature fruits are the largest and sweetest. Underground structures also change throughout the year.

Habitat also has a strong influence on plant growth. The leaves of plants from moist, shady sites are often larger, sweeter and more tender than those of plants on dry, sunny hillsides. Berries may be plump and juicy one year, when shrubs have had plenty of moisture, but they can become dry and wizened during a drought. Without the proper nutrients and environmental conditions, plants cannot grow and mature.

Finally, the genetic make-up of a species determines how the plant develops and how it responds to its environment. Wild plant populations tend to be much more variable than domestic crops, partly because of their wide range of habitats, but also because of their greater genetic variability. Humans have been planting and harvesting plants for millennia, repeatedly selecting and breeding plants with the most desirable characteristics. This process has produced many highly productive cultivars—trees with larger, sweeter fruits, potatoes with bigger tubers and sunflowers with larger, oilier seeds. These crop species are more productive, and they also produce a specific product each time they are planted. Wild plants are much less reliable.

Wild species have developed from a broader range of ancestors growing in many different environments, so their genetic make-up is much more variable than that of domestic cultivars. One population may produce

Salmonberry (*Rubus spectabilis*)

Red-osier dogwood (*Cornus sericea*)

sweet, juicy berries while the berries of another population are small and tart; one plant may have low concentrations of a toxin that is plentiful in its neighbour. This variability makes wild plants much more resilient to change. Although their lack of stability may seem to reduce their value as crop species, it is one of their most valuable features. Domestic crops often have few defences and must be protected from competition and predation. As fungi, weeds and insects continue to develop immunities to pesticides, we repeatedly return to wild plants for new repellents and, more recently, for pest-resistant genes for our crop plants.

Disclaimer

This book summarizes interesting, publicly available information about many plants in Canada. It is not intended as a "how-to" guide for living off the land. Rather, it is a guide for people wanting to discover the astonishing biodiversity of our useful plants and to connect to our cultural traditions, especially those of the First Nations. Only some of the most widely used species in Canada, and only some of their uses, are described and discussed. Self-medication with herbal medicines is not recommended. Use of plant medicines and consumption of wild foods should only be considered under guidance from an experienced healer/ elder/herbalist. As a field guide, the information presented here is limited, and further study of species of interest should be made using other botanical literature. No plant or plant extract should be consumed unless you are absolutely certain of its identity and toxicity and of your personal potential for allergic reactions. The authors and publisher are not responsible for the actions of the reader.

THE BERRIES

Pacific Crab Apple *Malus fusca*

Also called: western crab apple, Oregon crab apple • *Pyrus fusca*

Pacific crab apple (*M. fusca*)

This is western North America's only native species of apple. The berries of Pacific crab apple were an essential food item for coastal First Nations, and are still highly valued today. Indeed, these trees were widely managed and a carefully guarded resource that was often regarded as private property. Large quantities were harvested in the past, typically from late summer into autumn, often after the first frost. Another method of harvest was to pick the berries in whole bunches when they were still green and leave them to soften in baskets. They were either eaten fresh with eulachon oil or stored for winter. Because of their high acidity, this fruit keeps extremely well without processing, becoming sweeter and softer over time. Traditionally, they were placed, raw or cooked, in bentwood cedar boxes or large watertight baskets lined with skunk-cabbage leaves and covered with water and a layer of animal grease. Today they are preserved by canning, freezing or by being made into jelly, which has excellent flavour and a pretty, amber colour. Crab apples were sometimes an important part of ceremonial activities, such as potlatches or large feasts. Historically, boxes of Pacific crab apples were a common trade item and were also used as gifts at special events such as weddings. One ethnographical report states that around 1900 one box of crab apples preserved in water might cost as much as ten pairs of Hudson's Bay blankets, a significant value (for the time) of about $10! The fruit of Pacific crab apple is rich in pectin, so it can be added to low-pectin fruits when making jams and jellies. Pacific crab apple was also considered a fattening medicine and blood purifier. After a long day of hunting, the fruit was eaten to "kill poison in muscles."

EDIBILITY: highly edible

FRUIT: Small (10–15 mm), egg-shaped crab apples, forming green and ripening to yellow, orange or purplish-red colour, turning tan after a frost. Very tart and sour-tasting, but developing a sweetness and depth of flavour after a frost or storage that renders them quite delicious.

SEASON: Flowers late April, fruit ripening July to September, often staying on the tree well into winter if not eaten by wildlife.

> **WARNING:** *The bark and seeds of this tree contain cyanide compounds and should only be consumed under the guidance of a trained professional and with extreme caution.*

DESCRIPTION: Deciduous small tree or tall shrub, 2–12 m tall, scraggly in growth habit, with sharp spur-shoots. Bark rough, brown or greyish, older bark deeply fissured. Leaves alternate, 4–10 cm long, elliptic to lance-shaped, toothed and 1–3 irregular lobes, pointed at the end, dark green to yellow-green above and paler, slightly hairy below. Leaves turn yellow and orange in the fall. Flowers fragrant, white to pinkish apple blossoms with 5 showy petals, in flat-topped clusters. Fruits fleshy pomes ("crab apples"), 10–15 mm long, hanging in long-stemmed clusters. Found in estuary fringes, moist woods, wetlands, on streambanks and upper beaches along coastal areas, often growing in dense thickets.

Arbutus   *Arbutus menziesii*

Also called: Pacific madrone, madrone, madrona

Arbutus (*A. menziesii*)

Arbutus is the only broad-leaved evergreen tree native to Canada! Pacific First Nations people sometimes ate its berries. However, because of the berries' high tannin content, they are particularly astringent, especially if not fully ripe. The crushed berries can be made into a sweet cider or preserved as jellies. To store them over winter, they were first boiled or steamed, and then dried. They were then soaked in warm water before being eaten. Strung on thread, the berries make a natural Christmas garland that lasts well if stored properly from year to year. A cider made from the berries was employed to stimulate the appetite. Legend has it that the Saanich of southern Vancouver Island tied their canoes to arbutus trees following the Great Flood; to this day some Saanich people will not burn arbutus wood because of the service it provided. Traditionally, the leaves and scarlet berries were used to make necklaces and decorations.

EDIBILITY: edible

FRUIT: An orange-red berry with a rough, glandular surface, up to 1 cm wide, containing several seeds, growing in upright to drooping clusters. Taste described as bland, mealy, astringent, bitter, sweet.

SEASON: Flowers April to May. Ripens September to November.

DESCRIPTION: Small to medium, broadleaf, evergreen tree, to 30 m tall (but usually much shorter) with a wide, tropical-looking crown. Bark thin, smooth, reddish brown, peels in papery flakes and strips, with newly exposed surfaces (very soft to touch, like a chamois cloth) yellowish green, soon reddening, thickening on old trunks and breaking into many small, dark brown flakes. Leaves alternate, simple, thick and leathery, oval to elliptic, 7–15 cm long and 4–8 cm wide. Flowers greenish white to white (sometimes pinkish), 6–8 mm long, with a sweet honey-like fragrance, in drooping, branched clusters at the ends of the branchlets. Grows in well-drained soils in dry, open forests and on exposed, rocky bluffs near sea level along the Pacific coast in extreme southwestern BC. Southern BC is the most northern extent of its range, and it is rarely found more than 10 km inland or at heights above 300 m.

WILD GARDENING: *Arbutus is an extremely decorative tree that looks stunning, as if it belongs in the tropics rather than coastal BC. Its berries are a cheerful, bright red-orange colour and with its glossy evergreen leaves it is a good choice for the ornamental or wildlife garden in all seasons.*

Junipers *Juniperus* spp.

Common juniper (*J. communis*)

Some tribes cooked juniper berries into a mush and dried them in cakes for winter use. The berries were also dried whole and ground into a meal that was used to make mush and cakes. In times of famine, small pieces of the bitter bark or a few berries could be chewed to suppress hunger. Dried, roasted juniper berries have been ground and used as a coffee substitute

and teas were occasionally made from the stems, leaves and/or berries, but these concoctions were usually used as medicines rather than beverages. Juniper berries are well known for their use as a flavouring for gin, beer and other alcoholic drinks. Tricky Marys can be made by soaking juniper berries in tomato juice for a few days and then following your usual recipe for Bloody Marys, but omitting the gin. The taste is identical and the drink is non-alcoholic.

Juniper berries can be quite sweet by the end of their second summer on the plant or in the following spring, but they have a rather strong, "pitchy" flavour that some people find distasteful. They can be added as flavouring to meat dishes (recommended for venison and other wild game, veal and lamb), soups and stews, either whole, crushed or ground and used like pepper. Rocky Mountain juniper sprigs were also sometimes placed amongst dried salmon or other stored foods to protect these against attack from insects and flies.

Juniper berry tea has been used to aid digestion, stimulate appetite, relieve colic and water retention, treat diarrhea and heart, lung and kidney problems, prevent pregnancy, stop bleeding, reduce swelling and inflammation and calm hyperactivity. The berries were chewed to relieve cold symptoms, settle upset stomachs and increase appetite. Oil-of-juniper (made from the berries) was mixed with fat to make a salve that would protect wounds from irritation by flies. Juniper berries are reported to stimulate urination by irritating the kidneys and will give the urine a violet-like fragrance. They are also said to stimulate sweating, mucous secretion, production of hydrochloric acid in the stomach and contractions in the uterus and intestines. Some studies have shown juniper berries to lower blood sugar caused by adrenaline hyperglycemia, suggesting that they may be useful in the treatment of insulin-dependent diabetes. Juniper berries also have antiseptic qualities, and studies by the National Cancer Institute have shown that some

Common juniper (*J. communis*)

Common juniper (*J. communis*)

junipers contain antibiotic compounds that are active against tumours. Strong juniper tea has been used to sterilize needles and bandages, and during the Black Death in 14th-century Europe, doctors held a few berries in the mouth as they believed that this would prevent them from being infected by patients. During cholera epidemics in North America, some people drank and bathed in juniper tea to avoid infection. Juniper tea has been given to women in labour to speed delivery, and after the birth it was used as a cleansing, healing agent.

Juniper berries were sometimes dried on strings, smoked over a greasy fire and polished to make shiny black beads for necklaces. Some tribes also scattered berries to be used for necklaces on anthills. The ants would eat out the sweet centre, leaving a convenient hole for stringing. Smoke from the berries or branches of junipers has been used in religious ceremonies or to bring good luck (especially for

Rocky Mountain juniper
(*J. scopulorum*)

hunters) or protection from disease, evil spirits, witches, thunder, lightning and so on. The berries make a pleasant, aromatic addition to potpourris, and vapours from boiling juniper berries in water were used to purify and deodorize homes affected by sickness or death.

These plants are decorative, particularly in the winter months, and make a hardy and drought-tolerant addition to the ornamental garden. Junipers can be very long-lived, with some specimens recorded as old as 1500 years.

EDIBILITY: edible, but with caution

FRUIT: Small fleshy cones ("berries") are ripe when bluish-purple to bluish-green colour.

SEASON: Berries form from May to June on female plants only and mature the following year, but are present on the plant all year round.

Common juniper (*J. communis*)

Common juniper (*J. communis*)

Creeping juniper (*J. horizontalis*)

Common juniper (*J. communis*)

DESCRIPTION: Coniferous, evergreen shrubs or small trees, to 20 m tall, with some species creeping low on the ground. Leaves scale-like, opposite, 1–5 mm long, in rows, dark green to yellowy. Male plants produce yellow pollen on cones 5 mm long. Females bear small 5–9 mm-wide berries, first green and maturing to a bluish-purple colour. Grows on open, dry rocky areas and grasslands.

37

Rocky Mountain juniper (*J. scopulorum*)

Common juniper (*J. communis*) grows to 1 m tall but is normally lower than this. Growth habit is branching, prostrate, trailing, forming wide mats 1–3 m in size. Leaves are needle-like, dark green above, whitish below, prickly, 1.5 cm long, in whorls of 3. Bark reddish brown, scaly, thin, shredding. Grows on dry, open sites and forest edges, gravelly ridges and muskeg from lowland bogs to plains and subalpine zones. Also called: ground juniper.

Creeping juniper
(*J. horizontalis*)

Creeping juniper (*J. horizontalis*) is a low shrub (seldom over 25 cm tall) with trailing branches. Leaves scale-like, tiny, in 4 vertical rows, lying flat against the branch. Grows in dry, rocky soils in sterile pastures and fields.

Rocky Mountain juniper (*J. scopulorum*) is a rarer species, and grows to 15 m tall. Leaves opposite, 5–7 mm long, in 4 vertical rows, young leaves often needle-like, but mature leaves tiny and scale-like. Grows on dry, rocky ridges, open foothills, grasslands and bluffs. (These junipers growing in south-western BC and the Puget Sound have recently been recognized as a separate species, *J. maritima*).

Rocky Mountain juniper (*J. scopulorum*)

Hawthorns *Crataegus* spp.

Red hawthorn (*C. columbiana*)

The fruits, or haws, of this species are edible. The taste of the haws, however, can vary greatly depending on the species, particular tree, time of year and growing conditions. The haws are usually rather seedy, with the flavour described as a range of sweet, mealy, insipid, bitter, astringent or even tasteless. Frosts are known to increase the sweetness of the haws. Historically these berries were eaten fresh from the tree, or dried for winter use. They were also often an addition to pemmican. The cooked, mashed pulp (with the seeds removed) was dried and stored in cakes as a berry-bread, which could be added to soup or eaten with deer fat or marrow. Haws are rich in pectin and if boiled with sugar can be a useful aid in getting jams and jellies to set without a commercial pectin product. They can also be steeped to make a pleasing tea or cold drink. The haws of English hawthorn are commonly called "bread and butter berries" in the UK, likely because of the starchy and somewhat creamy texture of the fruit as you nibble it, especially after a frost.

Red hawthorn (*C. columbiana*)

Black hawthorn (*C. douglasii*)

Hawthorn flowers and fruits are famous in herbal medicine as heart tonics, though not all species are equally effective. Studies have supported the use of hawthorn extracts as a treatment for high blood pressure associated with a weak heart, angina pectoris (recurrent pain in the chest and left arm owing to a sudden lack of blood in the heart muscle) and arteriosclerosis (loss of elasticity and thickening of the artery walls). Hawthorn is believed to slow the heart rate and reduce blood pressure by dilating the large arteries supplying blood to the heart and by acting as a mild heart stimulant. However, hawthorn has a gradual, mild action and must be taken for extended periods to produce noticeable results. Hawthorn tea has also been used to treat kidney disease and nervous conditions such as insomnia.

Dark-coloured haws are especially high in flavonoids and have been steeped in hot water to make teas for strengthening connective tissues damaged by inflammation. The haws were sometimes eaten in moderate amounts to relieve diarrhea (some indigenous peoples considered them very constipating).

The scientific name "*Crataegus*" derives from the Greek *kratos*, which means "strength" and refers to the hard quality and durability of the wood. The common name "hawthorn" derives from the Old English word for a hedge, or "haw"; the species was historically planted and worked into hedgerows where its spiky thorns, branching nature and durable wood made it a formidable and lasting barrier.

English hawthorn (*C. monogyna*)

EDIBILITY: edible

FRUIT: Fruits are haws, hanging in bunches; small, pulpy, red to purplish pomes (tiny apples) containing 1–5 nutlets.

SEASON: Flowers May to June. Haws ripen late August to September.

DESCRIPTION: Deciduous shrubs or small trees growing 6–11 m tall with strong, straight thorns growing directly from younger branches. Leaves alternate, generally oval, with a wedge-shaped base. Flowers whitish, 5-petalled, sometimes unpleasant-smelling, forming showy, flat-topped clusters, from May to June.

Black hawthorn (*C. douglasii*) grows to 11 m tall, with 1–2 cm-long thorns. Leaves are toothed to shallowly lobed. The haws are 1 cm long, purplish black in colour. Grows in forest edges, thickets, streamsides and roadsides in lowland to montane zones, mostly in the southern half of the province. Also called: thorn apple.

English hawthorn (*C. monogyna*) grows 5–14 m tall, with thorns 1–2 cm long borne on younger stems. Bark is dull brown, sometimes with orange-shaded cracks. Leaves are dark on top, paler underneath, 2–4 cm long, deeply lobed sometimes to the midrib, obovate. Fruits are dark red in colour, 1 cm long, containing a single, hard seed. Introduced to BC but native to Europe, NW Africa and Western Asia. Also called: common hawthorn, oneseed hawthorn, May tree.

Red hawthorn (*C. columbiana*) grows to 6 m tall, with 4–7 cm-long thorns and dark red–coloured haws that are egg-shaped. Grows in open prairies, meadows, streambanks and forest edges in steppe and montane zones.

Red hawthorn (*C. columbiana*)

Suksdorf's hawthorn (*C. suksdorfii*) grows 1–12 m high, with thorns 8–12 mm. Bark is scaly, rough, pale grey-brown on younger branches, grey on older wood. Leaves 2.5–7.5 cm long, oblong to elliptic in shape, alternate, margins double-toothed, simple, pinnate. Fruit is black, shiny smooth. Similar to black hawthorn, differentiated by having significantly shorter thorns, leaves often less-lobed, and flowers with 20 stamens (very occasionally 15). Found in meadows, dry hillsides and riparian areas in southern BC, and tends to grow at a higher elevation than black hawthorn. Also called: Klamath hawthorn • *C. douglasii* var. *suksdorfii*.

Black hawthorn (*C. douglasii*)

Black hawthorn (*C. douglasii*)

43

Mountain Ash *Sorbus* spp.

Sitka mountain ash (*S. sitchensis*)

The bitter-tasting fruits of these trees are high in vitamin C and can be eaten raw, cooked or dried. In British Columbia, the Halkome'lem, Lillooet and Nlaka'pamux peoples are known to have consumed sitka mountain ash fruit and possibly those of western mountain ash, but most Canadian aboriginal groups considered them inedible. After picking, these berries were sometimes stored fresh underground for later use. They were also added to other more popular berries or used to marinate meat such as marmot or as a flavouring for salmon head soup (this is a particularly nutritious and delicious traditional food, so is well worth making if you're cooking a large salmon). The green berries are too bitter to eat, but the ripe fruit, mellowed by repeated freezing, is said to be tasty enough.

This species has been used to make jams, jellies, pies, ale and also bittersweet wine, and the fruit is also enjoyed cooked and sweetened. In northern Europe the berries, which can be quite mealy, were historically dried and ground into flour, which was fermented and used to make a strong liquor. A tea made of the berries is astringent, and has been used as a gargle for relieving sore throats and tonsillitis.

European mountain-ash fruit has been used medicinally to make teas for treating indigestion, hemorrhoids, diarrhea and problems with the urinary tract, gallbladder and heart. Some indigenous peoples rubbed the berries into their scalps to kill lice and treat dandruff. European mountain ash is a popular ornamental tree, and the native mountain ashes make attractive garden shrubs, easily propagated from seed sown in autumn. The scarlet fruit can persist throughout the winter and the bright clusters of fruit attract many birds.

EDIBILITY: edible, but not great

FRUIT: Fruits berry-like pomes, about 1 cm long, hanging in clusters.

SEASON: Blooms June to July. Fruits ripen August to September.

DESCRIPTION: Clumped, deciduous shrubs or trees with pinnately divided, sharply toothed leaves growing to 10 m. Bark smooth, brownish, with numerous lenticels (raised ridges that are actually breathing pores) on young bark, turning grey and rough with age. Leaves compound leaflets, alternate, 11–17 off each stem. Leaflets serrated, narrow, darker above, paler below. Flowers white, about 1 cm across, 5-petalled, forming flat-topped clusters, 9–15 cm wide, smelly. Grows in sun-dappled woods, rocky ridges and forest edges, preferring moist areas and part to full sun.

European mountain ash (*S. aucuparia*) is a widely planted ornamental tree, to 15 m tall, with white-hairy buds, leaf stems and leaves (underneath at least), and orange to red fruit. This Eurasian species is widely cultivated and just as widely escaped. Also called: rowan tree.

Western mountain ash (*S. scopulina*)

Western mountain ash (*S. scopulina*)

European mountain ash (*S. aucuparia*)

Sitka mountain ash (*S. sitchensis*) is a tall shrub, to 4 m tall, with rusty-hairy (but non-sticky) twigs and buds, 7–11 dull, bluish green, round-tipped leaflets without teeth near the base, and crimson to purplish fruit. Grows in foothill to subalpine zones.

Western mountain ash (*S. scopulina*) is a tall shrub, to 4 m tall, with sticky twigs and buds, 9–13 shiny green, pointed leaflets, with teeth almost to the base, and orange fruit. Grows in moist to wet open forests, glades and from streambanks to higher elevations. Also called: Greene's mountain ash.

Wild Roses *Rosa* spp.

Swamp rose (*R. pisocarpa*)

Most parts of rose shrubs are edible and the fruit (hips), which remain on the branches throughout winter, are available when many other species have finished for the season. These hips can be eaten fresh or dried and are most commonly used in tea, jam, jelly, syrup and wine. Usually only the fleshy outer layer is eaten (see Warning). Because they are so seedy, some indigenous peoples considered rose hips as famine food rather than regular fare.

Rose hips are rich in vitamins A, B, E and K and are one of our best native sources of vitamin C. Three hips can contain as much as a whole orange! During World War II, when oranges could not be imported, British and Scandinavian people collected hundreds of tonnes of rose hips to make a nutritional syrup. The vitamin C content of fresh hips varies greatly, but that of commercial "natural" rose hip products can fluctuate even more.

Rose petals have a delicate rose flavour with a hint of sweetness and may be eaten alone as a trail nibble, added to teas, jellies and wines, or candied. Adding a few rose petals to a regular salad instantly turns it into a gourmet conversation piece and guests are often surprised at how delicate and sweetly delicious the petals taste. Do not add commercial rose petals to salads, however, as these are often sprayed with chemicals.

Rose petals have been taken to relieve colic, heartburn, headaches and mouth sores. They were also ground and mixed with grease to make a salve for mouth sores or mixed with wine to make a medicine for relieving earaches, toothaches and uterine cramps. Dried rose petals have a lovely fragrance and are a common ingredient in potpourri. Rose sprigs were hung on cradleboards to keep ghosts away from babies, and on the walls of haunted houses and in graves to prevent the dead from howling. During pit cooking, the leaves of Nootka rose were placed under and over food to add flavour and prevent food from burning too close to the bottom of the pit. Hunters made a wash from Nootka rose branches to get rid of their human scent.

Some native roses can hybridize with each other, resulting in offspring that have mixed traits.

EDIBILITY: edible

FRUIT: Fruits scarlet to purplish, round to pear-shaped, berry-like hips, 1.5–3 cm long, with a fleshy outer layer enclosing many stiff-hairy achenes.

SEASON: Blooms June to August. Hips ripen August to September.

Nootka rose (*R. nutkana*)

Baldhip rose (*R. gymnocarpa*)

Prickly rose (*R. acicularis*)

DESCRIPTION: Thorny to prickly, deciduous shrubs, often spindly, 30 cm to 2 m tall. Spines generally straight (introduced rose species, of which there are a few naturalized in southern BC, tend to have curved spines). Leaves alternate, pinnately divided into about 5–7 oblong, toothed leaflets, generally odd in number. Flowers light pink to deep rose, 5-petalled, fragrant, usually growing at the tips of branches. Grows in a wide range of habitat: dry rocky slopes, forest edges, woodlands and clearings, roadsides and streamsides at mid- to low-level elevations.

Nootka rose (*R. nutkana*)

Arkansas rose (*R. arkansana*) is a small shrub, to 50 cm tall, with densely bristly stems that die back to near to base each fall. Prickles of various sizes, flowers clustered at end of branches, and apple-like hips 8–13 mm across. Grows in dry grassy slopes, prairies, banks and open

Arkansas rose (*R. arkansana*)

Baldhip rose (*R. gymnocarpa*)

Swamp rose (*R. pisocarpa*)

forests, present but not common in BC. Also called: prairie rose.

Baldhip rose (*R. gymnocarpa*) grows 50 cm to 2 m tall and has few to abundant soft prickles, with clusters of small flowers 2–3 cm wide, and 5–10 mm-wide hips. Hips have no attached sepal lobes; this is unique among our native species, and leaves the end of the hip "bald," hence the common name for this species. Grows in open forests and open areas in southern BC. Also called: dwarf rose.

Nootka rose (*R. nutkana*) is a small shrub, to 1.5 m tall, with well-developed thorns at its joints, and generally no prickles except occasionally on new wood. Flowers large (4–8 cm wide), mostly single, and 1.5–2 cm-long hips. Grows in moist, open areas (shorelines, forest edges, streambanks, roadsides) in southern BC.

WARNING: *The dry inner "seeds" (achenes) of the hips are not palatable and their fibreglass-like hairs can irritate the digestive tract and cause "itchy bum" if ingested. As kids we used to make a great old-fashioned itching powder by slicing a ripe hip in half and scraping out the seeds with these attached hairs. Spread this material to dry, then swirl it in a bowl, and the seeds will drop to the bottom. Skim off the fine, dry hairs—this is your itching powder, guaranteed to work. Although all members of the Rose family have cyanide-like compounds in their seeds, drying or cooking destroys the compounds.*

Baldhip rose (*R. gymnocarpa*)

Prairie rose (*R. woodsii*)

Prairie rose
(*R. woodsii*)

Prairie rose (*R. woodsii*) is a small shrub, 20–50 cm tall, with well-developed thorns at its joints, no small bristles or prickles on upper stems, small clusters of 3–5 cm-wide flowers followed by 6–10 mm-long hips. Grows in thickets, prairies and on riverbanks. Also called: Woods' rose.

Prickly rose (*R. acicularis*) grows to 1.5 m tall, with bristly, prickly branches and small clusters of 5–7 cm-wide flowers or 1–2 cm-long hips. Grows in open woods, thickets and on rocky slopes in BC. Prickly rose is Alberta's floral emblem.

Swamp rose (*R. pisocarpa*) grows 0.5–2 m tall, often with long, arching stems. Leaflets sharp-pointed, with paired prickles at leaf joints (like *R. nutkana*). Flowers small (< 4 cm), clustered, hips 6–12 mm long. Inhabits moist areas in the lower Fraser Valley and Vancouver Island. Also called: clustered wild rose, cluster rose.

Prickly rose (*R. acicularis*)

Prickly rose (*R. acicularis*)

Rosehip Jelly

Makes 8 x 1 cup jars

2 lbs whole rosehips · 2 lbs apples · 5 cups water · juice of 1 lemon
6 to 8 cloves · small cinnamon stick · white sugar

Carefully wash rosehips and apples. *While any ripe rosehips will work, in my experience those of the swamp rose have the most superior flavour. Slightly unripe apples work best for this recipe as they have a higher pectin content than ripe fruit does.* Core apples and chop roughly. Place the fruit in separate cooking pans with 2½ cups of water in each pan. Add lemon juice, cloves and cinnamon to the pan containing the rosehips. Bring both pans gently to the boil, then reduce heat and simmer until the fruit is soft and pulpy. Place the contents of both pans together in a jellybag and allow the juice to strain through overnight into a clean bowl. *If you want a perfectly clear jelly, do not press or squeeze the bag.*

In the morning, measure the strained liquid and allow for 2 cups of sugar to every 2½ cups of juice. Place the juice and sugar in a thick-bottomed cooking pan. *A thick-bottomed pan is important, because a thin-bottomed pan will get too hot and scald the jelly.* Bring to the boil, stirring and being careful to scrape the bottom of the pan, until the sugar is dissolved. Boil until setting point is reached (when you take some of the liquid on a wide-lipped spoon, blow on it to cool, then start to pour it off the side of the spoon and it gels together). Meanwhile, prepare 8 x 236 mL jars and lids (wash and sterilize jars and lids, and fill jars with boiling water; drain just before use).

Pour the hot jelly into clean, hot, sterilized jars. Seal the jars and place out of the sun to cool.

Prairie rose (*R. woodsii*)

Red Cherries *Prunus* spp.

Bitter cherry (*P. emarginata*)

These cherries may be eaten raw as a tart nibble, but the cooked or dried fruit is much sweeter and additional sugar further improves the flavour. The fruit can be cooked in pies, muffins, pancakes and other baking, or strained and made into jelly, syrup, juice, sauce or wine. It seldom contains enough natural pectin to make a firm jelly, however, so pectin must be added (see hawthorns or Pacific crab apple for a natural source rather than store-bought preparations). Although wild cherries are small compared to domestic varieties, they can be collected in large quantities. However, pitting such small fruits is a tedious job, especially since they are too tiny to use with a cherry-pitting tool.

Bitter cherry fruit (as the common name suggests) is bitter tasting and not that nice to eat without being sweetened, especially if not fully ripe. When in flower, the pin cherry tree is a dramatic and sweet-scented pleasure to behold so is well worth considering for the ornamental garden. Traditionally, pin cherry fruit were eaten fresh, cooked or dried and then powdered to store for winter use.

Wild cherries are worth growing in the ornamental or habitat garden as these are a favourite food for many mammals such as chipmunks, rabbits, mice, deer, elk and moose, and birds such as robins and grouse.

EDIBILITY: highly edible

FRUIT: Fleshy drupes (cherries) ranging in colour from red to blackish purple to black, with large stones (pits).

SEASON: Flowers April to June, fruits ripening July to August.

Bitter cherry (*P. emarginata*)

Bitter cherry (*P. emarginata*)

Bitter cherry (*P. emarginata*)

Pin cherry (*P. pensylvanica*)

DESCRIPTION: Deciduous shrubs or small trees growing 1–25 m tall. Trunk and branches reddish brown, often shiny, with raised horizontal pores (lenticels) prominent in stripes on the trunk and larger branches. Leaves smooth, finely toothed, sharp-tipped, 3–10 cm long. Flowers white or pinkish, about 1 cm across, 5-petalled, forming small, flat-topped clusters.

Bitter cherry (*P. emarginata*) grows to 15 m tall. Bark reddish brown or grey in colour. Leaves finely rounded at the tip, 3–8 cm long, oblong to oval, stalked, 1–2 small glands at base of leaf blade, blunt toothed with no prominent point (note the leaf shape, fruit colour and number of clustered fruit to differentiate this species from *P. pensylvanica*). Flowers 5–10 in flat-topped cluster,

Bitter cherry (*P. emarginata*)

10–15 cm across, white to pinkish. Fruit red to black cherries 8–12 mm long, ripening in August. Grows in moist, sparsely wooded areas along streambanks or cleared fields, hillsides and open or disturbed areas.

Pin cherry (*P. pensylvanica*) grows to 12 m tall. Leaves alternate, slender with long-tapering points, lance shaped, sharp toothed to 10 cm across. Flowers in clusters, 5–7 along twigs. Fruit bright red cherries, 4–8 mm long, thin sour flesh, growing in elongated clusters, 10+ per bunch. Grows in moist thickets, woods, riverbanks, forest edges, disturbed areas and clearings from sea level to subalpine zones. Also called: bird cherry, fire cherry, Pennsylvania cherry.

Pin cherry (*P. pensylvanica*)

Pin cherry (*P. pensylvanica*)

Chokecherry *Prunus virginiana*

Also called: wild cherry

Chokecherry (*P. virginiana*)

Chokecherries were among the most important and widely used berries by First Nations across Canada. In BC these fruits were highly regarded, especially amongst Interior First Peoples. They were collected after a frost (which makes them *much* sweeter) and were dried or cooked, often as an addition to pemmican or stews. Large quantities were gathered, pulverized with rocks, formed into patties about 15 cm in diameter and 2 cm thick and dried for winter use. They were most commonly dried with the pits intact (a process that destroys the toxic hydrocyanic acid in the pits) and could also be stored when picked as branches for several months if kept in a cool, dry place. Today, chokecherries are used to make beautiful-coloured jelly, syrup, sauce and beer as well as wine.

The raw cherries are sour and astringent, particularly if they are not fully ripe, so they cause a puckering or choking sensation when they are eaten—hence the common name "chokecherry." One unimpressed early European traveler in 1634 is reported to have written that "chokecherries so furre the mouthe that the tongue will cleave the roofe, and the throate wax hoarse"! After they have been cooked or dried, however, they are much sweeter and lose their characteristic astringency.

Dried, powdered cherry flesh was taken to improve appetite and relieve diarrhea and bloody discharge of the bowels. For the Okanagan–Colville First Nations, the ripening chokecherry fruits signaled that the spring salmon were coming up the river to spawn (in scientific circles this is called a "phenological indicator," where the appearance or life stage of one organism signifies a corresponding stage in another organism). Dried chokecherries were an important trade item for some Canadian indigenous peoples and the Interior Shuswap people also used the berries as a paint for creating pictographs.

EDIBILITY: highly edible (when fully ripe, after a frost, or sweetened)

FRUIT: Red, black to mahogany-coloured, shiny, growing in heavy and generous trusses. Some reports indicate that the red fruit have a nicer flavour than the darker-coloured ones.

SEASON: Flowers May to June. Ripening August to September.

DESCRIPTION: Deciduous shrub or most often small tree growing to 8 m tall. Bark smooth, greyish marked with small horizontal slits (slightly raised pores called lenticels). Leaves alternate, 3–10 cm long, broadly oval, finely sharp-toothed, with 2–3 prominent glands near the stalk tip. Flowers creamy white, 10–12 mm across,

5-petalled, forming bottlebrush-like clusters 5–15 cm long, in May to June, producing hanging clusters of 6–12 mm wide, dark red to black cherries. Chokecherry grows in deciduous woods, open sites, streams and forest edges, particularly in the dry Interior of the province.

WARNING: *Like other species of* Prunus *and* Pyrus, *all parts of the chokecherry (except the flesh of the fruit) contain cyanide-producing glycocides. There are reports of children dying after eating large amounts of fresh chokecherries without removing the stones. Cooking or drying the seeds, however, appears to destroy most of the glycocides. Chokecherry leaves and twigs are poisonous to animals.*

Indian-plum · *Oemleria cerasiformis*

Also called: bird plum, oso-berry, skunk bush · *Osmaronia cerasiformis, Nuttallia cerasiformis*

Indian-plum (*O. cerasiformis*)

The berries of Indian-plum were eaten fresh or dried by some BC First Nations, but in small quantities. The fruit has only a thin layer of flesh covering the large single seed, which makes it rather labour-intensive to prepare or consume. The fruits are very bitter before they become fully ripe and some First Nations referred to Indian-plum as "choke cherry" because of this. Reports claim that it tastes best when changing from red to purple in colour. This fruit was eaten fresh, dried or cooked, typically at family meals or at large feasts, and was also sometimes covered with oil and stored in cedar boxes for winter eating.

Ground squirrels and other rodents, birds, deer, foxes and coyotes also eat this fruit. Indian-plum is one of our first native species to bloom in the spring, and the flowers are an important early nectar source for bees and other insects.

The species name, *cerasiformis*, means "cherry-like." Another common name for Indian-plum is oso-berry; "oso" means "bear" as these animals are known to relish the berries.

EDIBILITY: edible

FRUIT: Fruit a fleshy drupe resembling bunches of small plums, with a mild but distinctive plum flavour, about 1 cm long, green then peach-coloured when unripe, bluish black when ripe, with a large pit, borne on a red stem.

SEASON: Flowers April to May. Fruits ripen in June.

DESCRIPTION: Deciduous shrub or small tree, 1–5 m tall. Bark smooth and grey, reddish to purplish on new twigs. Leaves alternate, blades 5–12 cm long, lance-shaped to oblong, smell like cucumber or watermelon rind when crushed, light green and smooth above, paler beneath. Flowers greenish white, 1 cm across, male and female flowers similar-looking but borne on separate plants, generally appearing before leaves, in clusters hanging from leaf axils, strong smelling (male flowers are said to smell like cat urine, whereas female flowers smell like cucumber or watermelon rind). Found at low elevations in dry to moist, open woods, along rivers and streambanks and clearings in extreme southwestern BC.

Smooth Sumac *Rhus glabra*

Smooth sumac (*R. glabra*)

Sumac's showy, red fruit clusters are beautiful to look at and can be made into a refreshing and pretty, pink or rose-coloured drink that has a lemon-like flavour. Crush the berries then soak the mash in cold water before straining to remove the fine hairs from the fruit and other debris. The juice is best sweetened with sugar and served cold. Boiling fruit in hot water, however, releases tannins and produces a bitter-tasting liquid. The fruits have also been used to make jellies or lemon pies. The tangy lemon flavour of sumac fruit (which really comes from the hairs covering the seeds) is a common ingredient in some Middle Eastern dishes.

The fruits were boiled to make a wash to stop bleeding after childbirth. The berries, steeped in hot water, made a medicinal tea for treating diabetes, bowel problems and fevers. This tea was also used as a wash for ringworm, ulcers and skin diseases such as eczema. When chewed as a trail nibble, sumac fruits relieve thirst and leave a pleasant taste in the mouth.

Sumac is a very decorative and hardy species that provides an interesting fall and winter garden display. It does tend to sucker, though, so can get invasive in the garden if not kept in check.

EDIBILITY: edible

FRUIT: Fruits reddish, densely hairy, berry-like drupes, 4–5 mm long, in persistent, fuzzy clusters.

SEASON: Flowers May to July. Fruits ripen July to August, often remaining through the winter.

DESCRIPTION: Deciduous shrub or small tree, 1–3 m tall, usually forming thickets. Twigs and leaves are hairless; buds have whitish hairs. Branches exude milky juice when broken. Leaves pinnately divided into 11–31 lance-shaped, 5–12 cm long, toothed leaflets, bright red in autumn. Flowers cream-coloured to greenish yellow, about 3 mm across, with 5 fuzzy petals, forming dense, pyramid-shaped, 10–25 cm-long clusters. Grows in dry forest openings, prairies, fencerows, roadsides and burned areas in southern BC.

Indian Lemonade

Makes 8½ cups

Enhance this beautiful pink "lemonade" by adding ice cubes (or frozen blueberries) and green mint sprigs.

3 cups dried and crumbled sumac flower spikes
8 cups water • sugar to taste

Pick through the dried flower spikes to remove any dirt or debris. Crumble the red "berries" off the main spike and place them in a jug. Pour the cold water over the berries, mash the mixture with a wooden spoon or potato masher, then let sit for at least an hour. *Do not heat this mixture because it will alter the taste of the sumac.* Strain the liquid through a cheesecloth, jellybag or fine-mesh sieve, and add sugar to taste.

Blackberries *Rubus* spp.

Himalayan blackberry (*R. armeniacus*)

Blackberries and their relatives (raspberries, salmonberry, thimbleberry and cloudberry) are all closely related members of the genus *Rubus*. The best way to distinguish blackberries from raspberries is by looking at their fruits: if they are hollow like a thimble, they are raspberries, and if they have a solid core, they are blackberries.

Blackberries were traditionally gathered by indigenous peoples and are still widely enjoyed today. They were typically gathered in large quantities and eaten both fresh and stored (usually dried, either alone or with other fruit) for winter. A traditional method of eating the berries was to combine them with other berries or with oil (sometimes whipped) and meat. Blackberries today are enjoyed in many different ways: on their own, with cream or yogurt and sugar, in pies and sauces, as jam or jelly or as drinks such as cordial, juice or wine.

Our native trailing blackberry is the first to ripen in early July, followed by the introduced Himalayan blackberry and Allegheny blackberry, then the cutleaf blackberry which starts to

Himalayan blackberry (*R. armeniacus*)

SEASON: Flowers May to July (sometimes into August in moist, shady or cool spots). Fruits ripen June to September.

DESCRIPTION: Prickly, perennial shrubs, often arching or trailing, branches 50 cm to 5 m long. Leaves alternate, compound. Flowers white to pinkish, 5-petalled. Fruit fall from the shrub with the fleshy receptacle intact (i.e., the blackberries have a solid core).

Allegheny blackberry (*R. allegheniensis*)

Himalayan blackberry (*R. armeniacus*)

ripen significantly later in August. Blackberries are widely cultivated across Canada for their delicious fruits; modern thornless cultivars are readily available for the home gardener. Blackberry-raspberry crosses, such as loganberries and boysenberries, are extremely flavourful and should be more popular and widely known than they are.

In an origin story for trailing blackberry, some Coast Salish First Nations on BC's south coast recount that a jealous husband chased a woman up a tree. As the blood of the woman fell from the tree and reached the ground, the drops turned into blackberries. A purification rite of the same First Nation involved scrubbing trailing blackberry stems across a person's body before spirit dancing.

The word "bramble" comes from the Old English "*braembel*" or "*brom*," which means "thorny shrub."

EDIBILITY: highly edible

FRUIT: Juicy red to black drupelets aggregated into clusters.

65

Allegheny blackberry (*R. allegheniensis*) is a medium-to-large erect shrub, to 3 m tall. Stems and leaves glandular-hairy; deciduous leaves usually have 5 leaflets arranged palmately (like the fingers on a hand). Native to Eastern Canada, and introduced to BC where it has naturalized and can be found in dry thickets, clearings and woodland margins only around the lower Fraser Valley.

Cutleaf blackberry (*R. laciniatus*) grows 2–10 m or more, often as a thicket, generally evergreen. Stems 3–10 mm, first ascending and arching, then sprawling, trailing, rooting at stem tips. Prickles flattened, stout, hooked. Leaflets 5, roughly egg-shaped, deeply lobed or toothed, tip sharp-pointed, smooth green, slightly hairy above, greyish green hairy below. Flowers pinkish to white, petals spreading, 3-lobed at tip. Fruits drupelets 1–1.5 cm, red ripening to a very glossy black, globe- to egg-shaped, clustered. Fruit has a distinctive taste, is much firmer in consistency than the other species, ripens later, and has a more pronounced core. Also called: evergreen blackberry.

Himalayan blackberry (*R. armeniacus*) is a large shrub, to 5 m tall, similar to Allegheny blackberry, but its leaves can be evergreen, and its leaves and stems lack glandular hairs. This introduced species grows in open, disturbed sites in southwestern coastal BC. It is considered an invasive species, particularly in the endangered Garry oak meadow habitat of southern Vancouver Island. Also called: *R. procerus, R. discolor.*

Trailing blackberry (*R. ursinus*) is a trailing shrub with arching stems, less than 50 cm tall but 5 m or more long, rooting at the tip of the shoot. Leaflets 3, deciduous. Found in thickets, dry open forests, fields, streamsides and disturbed sites in southwestern BC. Also called: dewberry, Pacific blackberry.

Cutleaf blackberry (*R. laciniatus*)

Trailing blackberry (*R. ursinus*)

Blackberry Syrup

Makes 3 x 1 cup jars

This fruity syrup makes a delicious warm or cold drink and is recommended for relieving the symptoms of the common cold. Add 1 Tbsp to 1 cup of hot water.

1 lb blackberries · 1 cup white wine vinegar · 1 cup sugar · 4 Tbsp honey

Place the clean fruit into a glass jar and pour the vinegar over top. Leave to stand for at least 24 hours, stirring and crushing the fruit regularly to extract the juices. Strain the liquid into a large saucepan and bring to the boil. Add the sugar, stirring until it's all dissolved. Add honey, stir well. Bring back to the boil, and boil hard for 5 minutes. Allow to cool completely. Pour into sterilized jars while still hot, and seal. Alternatively, you can pour the cooled liquid into ice cube trays in the freezer.

TIP

If you are pouring very hot liquid into a sterilized glass jar that has cooled, the sudden heat can cause the jar to crack. Avoid this problem by first pouring in a few tablespoons of the hot liquid and waiting 10 seconds for the heat to spread, then filling the rest of the jar.

Berry Blackberry Cordial

Makes approximately 4 x 1 cup jars if 8 cups of fruit are used

up to 8 cups freshly picked blackberries (or other juicy berries such as raspberries or thimbleberries, or a combination of berries) · white vinegar · sugar

Carefully pick through the fruit to remove any debris or insects. Be particularly wary of stink bugs, which are about 1 cm in size, green to brownish in colour, flat-backed with a hard carapace, and emit a rank stench if bitten into: they will ruin the entire batch of cordial!

Place the berries in a large glass jar and crush somewhat firmly with a potato masher. Pour enough white vinegar into the jar to just barely cover the fruit mash (roughly an 8:1 ratio). Stir vigorously, put a firm lid on the jar, then let it sit somewhere warm out of direct sunlight for 1 week, stirring once a day.

After a week, strain the mixture overnight through a jellybag. *Resist squeezing it or you will push solids through the bag, resulting in a cloudy end-product with sediment.* The leftover fruit mash can be used in muffins or pancakes.

Measure out the resulting juice into a thick-bottomed saucepan and add 1 cup white granulated sugar for every 1 cup of juice. Slowly bring to the boil to fully dissolve the sugar. Let cool and place in washed, sterilized Mason-type jars for storage. Other glass containers such as maple syrup bottles with rubber-sealed tops also work well.

To make the cordial, mix the concentrate in a 6:1 ratio with cold water. Garnish with a sprig of fresh mint or some frozen whole berries.

Raspberries *Rubus* spp.

Arctic dwarf raspberry (*R. arcticus*)

Raspberries and relatives (salmonberry, thimbleberry and cloudberry) are all closely related members of the genus *Rubus*. The best way to distinguish these berries from blackberries is by looking at their fruits: if they are hollow like a thimble, they are raspberries (or relatives), and if they have a solid core, they are blackberries.

Wild raspberries are one of our most delicious native berries and are fabulous fresh from the branch or made into pies, cakes, puddings, cobblers, jams, jellies, juices and wines. Since the cupped fruit clusters drop from the receptacle when ripe, these fruits are soft and easily crush to a juicy mess when gathered. Raspberries were a popular and valuable food of indigenous peoples and were often gathered and processed into dried cakes either alone or with other berries such as salal for winter use. These cakes were reconstituted by boiling, or eaten as an accompaniment to dried meat or fish. The Coast Salish mixed western black raspberries with black twinberry, salal and wild red raspberry to make a purple stain. Fresh or dried leaves of this species have been used to make tea, and the flowers make a pretty addition to salads. Although trailing wild raspberry fruit are delicious and juicy, they were not traditionally gathered in quantity because of their small size and the difficulty in picking them.

Raspberry leaf tea and raspberry juice boiled with sugar have been gargled to treat mouth and throat inflammations.

Trailing wild raspberry (*R. pedatus*)

Western black raspberry (*R. leucodermis*)

Arctic dwarf raspberry (*R. arcticus*)

EDIBILITY: highly edible

FRUIT: Fruits juicy red to black drupelets aggregated into clusters that fall from the shrub without the fleshy receptacle (raspberries and relatives have a hollow core).

SEASON: Flowers June to July. Berries ripen July and August.

DESCRIPTION: Armed or unarmed, perennial shrubs or herbs, 15 cm to 4 m tall. Leaves deciduous, lobed or compound (divided into leaflets). Flowers white to pink.

Arctic dwarf raspberry (*R. arcticus*) is a low, herbaceous plant (sometimes a bit woody at the base), to 15 cm tall, with typically 3 leaflets, no prickles or bristles, pink to reddish flowers, and deep red to dark purple fruits. Found in bogs, wet meadows and tundra. Also called: nagoonberry, Arctic raspberry • *R. acaulis.*

Dwarf bramble (*R. lasiococcus*) has slender trailing stems to 2 m long, with short erect flowering branches to 10 cm tall. Leaves 1–3 per erect branch, heart- to kidney-shaped, double saw-toothed, 2–6 cm wide. This species is very similar to *R. pubescens* (both have leaves which appear 3-lobed, rather than the 5-lobed leaves of *R. pedatus*). Fruits drupelets, to 1 cm wide, few to several in a cluster. Has a very limited range in extreme southern BC. Also called: rough-fruit berry.

Trailing raspberry (*R. pubescens*)

Trailing raspberry (*R. pubescens*) is a slender trailing, soft-hairy shrub, unarmed, to 50 cm tall and often more than 1 m long (vegetative stems are ascending at first, then reclining), rooting where the nodes touch the ground. Leaflets alternate, long-stalked, greenish and smooth hairy above, paler beneath, with 3–5 lobes. Flowers 5-lobed, 1–3 on erect shoots, 15–50 cm, blossoms white, rarely pinkish. Fruits dark red drupelets, to 1 cm, smooth, several, not easily separating from spongy receptacle. Found on damp slopes, rocky shores and low thickets. Also called: dewberry.

Western black raspberry (*R. leucodermis*)

Wild red raspberry (*R. idaeus*)

Trailing wild raspberry (*R. pedatus*) is a trailing herb from long, creeping stems to 1 m long, rooting at the nodes and producing short erect stems bearing flowers and 1–3 leaflets, usually less than 10 cm tall. Leaflets alternate, 3–5, coarsely toothed, no prickles or bristles. Flowers solitary, on slender stalks, white, petals spreading or bent backwards. Fruits dark or bright red clusters of flavourful and juicy drupelets (raspberries), sometimes only 1 drupelet per fruit. Found in low to subalpine elevations in bog forests, streambanks and moist, mossy forests. Also called: strawberry leaf raspberry, creeping raspberry, five-leaved bramble.

Western black raspberry (*R. leucodermis*) has long, arching branches to 2 m long. Stems glaucous-grey, highly prickly. Leaves alternate, crinkly, deciduous, shiny white beneath. Leaf stalks stout, prickly, broad-based, and usually hooked. Flowers white to pink in colour, clustered in leaf axils or at terminal end of stalk. Fruit "hairy" raspberries, to 1 cm, red (when unripe) to dark purple-black. Inhabits thickets, ravines, disturbed areas, forest edges and open woods. Also called: blackcap, whitebark raspberry.

Wild red raspberry (*R. idaeus*) is an erect shrub, to 2 m tall, growing in thickets as it spreads by underground rhizomes. Fruit bright red, virtually identical to the domesticated raspberry, but smaller. Most common east of the Cascade and Coast Mountain ranges. Grows in thickets, open woods, fields and on rocky hillsides across Canada. Also called: American red raspberry • *R. strigosis*.

Wild Berry Dressing

Makes about 2 cups

This dressing keeps well in the fridge for up to 10 days.

1 cup mixed tangy wild berries such as raspberries, thimbleberries or blackberries
½ cup olive oil • ¼ cup apple cider vinegar
1 tsp sugar • 2 cloves crushed garlic • 1 tsp salt

Crush the berries, then mix with all the remaining ingredients in a small jam jar. Screw on the lid tightly and shake vigorously.

Wild red raspberry (*R. idaeus*)

Trailing wild raspberry (*R. pedatus*)

Cloudberry *Rubus chamaemorus*

Also called: bake-apple

Cloudberry (*R. chamaemorus*)

Cloudberry was historically, and is still, a principal food for northern indigenous peoples; these juicy berries are delicious, with a distinctively tart taste that some reports say is acquired. Cloudberries have twice as much vitamin C by volume as an orange and were an important food against scurvy for First Nations and early northern immigrants. Traditionally, these summer fruits were stored in seal pokes (containers made by cleaning, inflating and drying a complete sealskin), wooden barrels or underground caches in cold water or oil, with other berries or with edible greens. The Latin for this species derives from the Greek words "*chamai*" meaning "on the ground" and "*Moros*" meaning "mulberry."

EDIBILITY: highly edible

FRUIT: Fruit raspberry-like in appearance, each made up of 5–25 drupelets, amber to yellow when mature.

SEASON: Flowers May to June. Fruits ripen in August.

DESCRIPTION: A low, unbranched herb, to 25 cm tall, with 1–3 leaves per stem. Leaves round to kidney-shaped (not divided into leaflets), shallowly 5- to 7-lobed, no prickles or bristles. Flowers single, white, at end of stem, the male and female flowers on different plants. Found in peat bogs and peaty forests at northern latitudes.

Berry Fruit Leather

Makes 1 baking sheet of fruit leather

4 cups crushed berries (all one kind or a mix) · 2 cups apple sauce · ½ cup sugar

Mix the berries and sugar together in a pot on medium heat until the sugar is dissolved. Put the mixture through a food mill to remove any stems or seeds, then add the apple sauce and stir until well mixed. Grease a rimmed baking sheet and pour mixture in. Use a spatula to spread the mixture to an even thickness on the baking sheet, because the fruit leather will not dry evenly otherwise. Place in a food dehydrator or an oven at 150° F until firm to the touch and dry enough to peel off. Remove from the dehydrator or oven and let cool. Use scissors to cut the leather into strips. Cool the strips and store in an airtight container or Ziploc® bag.

Thimbleberry   *Rubus parviflorus*

Thimbleberry (*R. parviflorus*)

Thimbleberry is one of the most delicious native berries you will encounter in BC (and beyond!) and was highly regarded by all First Nations in its range. The fruit is easy to pick as it can grow in large clusters and appears on the plant as a bright red treasure amid soft, maple-like leaves (yay, no sharp prickles or spines!). The taste is somewhat like a raspberry, but more intense and flavourful with a sharper "tang." Once you've had thimbleberry pie, jam or tarts you will never go back to its poorer cousin—the raspberry!

The fruit, which is rather coarse and not overly juicy, dries and keeps well. This species can also be gathered by cutting the stems of the unripe fruit, which will ripen later in storage.

Traditionally, these fruit were gathered, mashed either alone or with other seasonal fruits such as salal, and dried into cakes for winter use or trade. Tender shoots of this plant were traditionally harvested and peeled in the early spring as a green vegetable.

The large, maple-like leaves of thimbleberry served many purposes for some indigenous peoples in BC. They were used to whip soapberries, wipe the slime from fish, line and cover berry baskets and dry other kinds of berries. To make a temporary berry container, pick a leaf and then snap off the stem. Fold the outer soft leaf edges together to form a funnel shape (the stem is at the narrow, bottom edge of this funnel, the leaf tips forming the wider top brim), then use the stem to prick through the two leaf folds where they overlap and "sew" the funnel together. If you still have a small hole at the bottom of your funnel, line this with part of another leaf. If you're out in the woods and have forgotten your toilet paper, thimbleberry leaves are soft and tough and make an excellent substitute.

EDIBILITY: highly edible

FRUIT: Bright red, shallowly domed (like a thimble), raspberry-like hairy drupelets, in clusters held above the leaves.

SEASON: Flowers April to May. Fruits ripen July to August.

DESCRIPTION: Erect shrub, 0.5–3 m tall, main-stemmed, with no prickles or spines, spreading by underground rhizomes and forming dense thickets. Bark light brown and shredding on mature stems, green on newer stems. Leaves large (up to 15 cm), soft, fuzzy, maple-like, palmate, 3–7 lobes, alternate, toothed around margins, fine hairs above and below. Flowers white, 5-petalled, large to 4 cm, long-stemmed in clusters of 3–11, terminally clustered. Found in moist open sites such as road edges, shorelines and riverbanks and open forests at low- to mid-elevations.

Wild Berry Juice

Makes approximately 4½ cups

3 cups any sweet berries such as blueberries, thimbleberries or blackberries
2 cups water • sugar to taste

Pick over berries to remove any debris and place them in a saucepan with the water. Mash the mixture with a wooden spoon on potato masher, then simmer until berries are soft. Strain the mixture through a jellybag, fine-mesh sieve or cheesecloth, add sugar to taste, then let the juice cool before serving.

Salmonberry *Rubus spectabilis*

Salmonberry (*R. spectabilis*)

Salmonberry fruits and shoots were popular foods for coastal indigenous peoples in BC. They were almost always eaten fresh, sometimes at feasts, because they are very juicy and do not preserve well. The shoots are still peeled and enjoyed raw, steamed or boiled as an early spring vegetable.

The flowers of this species are a magnificent rosy red to reddish-purple colour and make an excellent addition to the ornamental garden. Beware, though, of this plant's habit of spreading through underground rhizomes as it can become invasive.

These are one of the first flowers to come out in the spring; in fact, the Haisla consider the arrival of the first salmonberry flower to signify the arrival of the new year. In a Haisla tradition, the first pink blossoms are spread about the house to celebrate the arrival of the new season. Berries were traditionally eaten with grease, salmon, dried salmon eggs, and sometimes mixed with other berries, sugar and a little water. When salmonberries bloom, this indicates to the Haisla and Hanaaksiala peoples that it is time to travel to their places to pick and process edible seaweed.

These berries are one of our first to ripen on the coast, and are a tantalizing harbinger of the other summer fruits yet to come.

Swainson's thrush is called the "salmonberry bird" in some coastal languages because it is associated with ripening salmonberry in northwest coast native mythology.

EDIBILITY: highly edible

FRUIT: Large, raspberry-like fruits, ranging when ripe in a remarkable palette from gold to ruby red to purplish black. Flavour varies greatly depending on plant and individual location.

SEASON: Flowers bloom February to March. Fruits ripen May to June.

DESCRIPTION: Robust, erect shrub, to 4 m tall, growing in thickets from spreading rhizomes. Bark reddish brown to yellowish, shredding. Prickles most present on younger branches, sloughing off as the bark matures. Leaves alternate, deciduous, sharply toothed, normally with 3 leaflets. Flowers pink to reddish, to 4 cm, 1–2, rarely up to 4, on short stems. Found in moist to wet forests, swampy areas, disturbed areas and streambanks along coastal BC.

Oregon-grapes *Mahonia* spp.

Tall Oregon-grape (*M. aquifolium*)

These tart, juicy berries can be eaten raw but can be rather sour and intense, so are more commonly used to make jelly, jam or wine. A frost increases the fructose content of the berries, thereby making them sweeter and more palatable for fresh eating. Mashed with sugar and served with milk or cream, they make a tasty dessert. A refreshing drink can be made with mashed berries, sugar and water—the sweetened juice tastes much like Concord grape juice. Berry production can vary greatly from year to year, and the fruits are sometimes rendered inedible by grub infestations, so the eater should be wary of the potential for extra protein! The plants of this species contain the alkaloids berberine, berbamine, isocorydin and oxyacanthine, which stimulate involuntary muscles. The crushed plants and roots have antioxidant, antiseptic and antibacterial properties. They were used to make medicinal teas, poultices and powders for treating gonorrhea and syphilis and for healing wounds and scorpion stings. Boiled, shredded root bark produces a beautiful, brilliant yellow dye.

Tall Oregon-grape is a decorative, tough, spiky, drought-tolerant and hardy plant that is often planted as an ornamental. It produces masses of bright yellow, scented flowers adored by pollinators early in the spring, followed by tresses of decorative purple berries loved by birds, and eventually a striking and lasting fall display of red leaves. Its sharp spiky leaves also make it a good plant in areas where you may want to discourage pets or people from exploring.

Dull Oregon-grape (*M. nervosa*)

Tall Oregon-grape (*M. aquifolium*)

Dull Oregon-grape (*M. nervosa*)

EDIBILITY: highly edible (see Warning)

FRUIT: Fruits juicy, grape-like berries, tart and sour tasting, about 1 cm long, purplish blue with a whitish bloom.

SEASON: Flowers April to June. Fruits ripen August to September.

DESCRIPTION: Perennial, evergreen shrubs, to 3 m tall. Outer bark rough, brown to greyish, and with a striking canary/mustard yellow inner layer when scraped. Leaves leathery, holly-like, pinnately divided into spiny-edged leaflets, dark glossy green, turning red or purple in winter. Flowers yellow, about 1 cm across, forming elongated whorled clusters. Oregon-grape is the state flower of Oregon.

Dull Oregon-grape (*M. nervosa*)

Creeping Oregon-grape (*M. repens*)

Creeping Oregon-grape (*M. repens*)

Creeping Oregon-grape (*M. repens*) is a creeping broadleaf evergreen shrub 10–20 cm tall, spreading by both underground rhizomes (root-bearing stems) and surface suckers. Leaves 3–7 dull or shiny leaflets, bluish green with a white bloom underneath, alternate. Grows in open forests, scrublands and grasslands at low to montane elevations in southern BC. Also called: holly grape • *Berberis repens.*

Dull Oregon-grape (*M. nervosa*) grows 10–80 cm tall. Leaves are leaflets, dull rather than shiny, whitish bloom underneath, in sets of 9–19 pairs, opposite from a common stem (botanists call this form "pinnate," from the Latin for "feather," *pinna*). Grows in moist to dry forests and open slopes at low to montane elevations in southwestern BC. Also called: Cascade Oregon-grape, Cascade barberry • *Berberis nervosa.*

Tall Oregon-grape (*M. aquifolium*) grows 50 cm to 3 m tall. Leaves 5–11 leaflets, shiny above and not whitened beneath, prominent central vein on each. Flowers grow both in upright clusters and in hanging racemes. Grows in dry forests at low to montane elevations in southern BC. Also called: Oregon-grape, holly-leaved barberry • *Berberis aquifolium*.

WARNING: *High doses of Oregon-grape can cause nosebleeds, skin and eye irritation, shortness of breath, sluggishness, diarrhea, vomiting, kidney inflammation and even death. Pregnant women should not use this plant, because it may stimulate the uterus.*

Tall Oregon-grape (*M. aquifolium*)

Salal and Oregon-grape Jelly

Makes 16 x 1 cup jars

8 cups salal berries • 8 cups Oregon-grape berries • ¼ cup lemon juice
1 packet powdered pectin • 5 cups sugar

Place salal and Oregon-grape berries in a thick-bottomed saucepan on medium heat. Crush and stir the berries and simmer until the juice is released, about 10 minutes. Strain through a cheesecloth or fine-mesh sieve to separate the juice. *Do not squeeze the cloth or force the mix through the sieve because it will cause sediments to run into the juice, resulting in a cloudy jelly.*

Measure out 4 cups of the resulting juice into a thick-bottomed saucepan. Add the lemon juice and pectin, stirring until the pectin is thoroughly dissolved. Add the sugar, stirring constantly, and bring to a rolling boil. Hold the mixture at the boil for 3 minutes, being careful to stir the bottom so that the jelly does not stick or burn.

Meanwhile, prepare 16 x 236 mL jars and lids (wash and sterilize jars and lids, and fill jars with boiling water; drain just before use).

Remove from heat, skim off any foam (the impurities coming out of the liquid) and pour into the hot, sterilized jars. Carefully wipe the jar edges to ensure they are clean and dry, then place the lids on and tighten the metal screw bands. Place jars in a cool area. You will know that the jars have sealed when you hear the snap lids go "pop."

Currants *Ribes* spp.

Red flowering currant (*R. sanguineum*)

Currants were eaten by many First Nations, and are common and widespread throughout BC. While all are considered edible, some are tastier than others, and some (such as wax currant and sticky currant) are considered emetic in large quantities and are best avoided. Currants are high in pectin and make excellent jams and jellies—either alone or mixed with other fruit—that are delicious with meat, fish, bannock or toast. Historically, currants were also mixed with other berries and used to flavour liqueurs or fermented to make delicious wines, but raw currants tend to be very tart.

Wax currants have been described as tasteless, dry and seedy, bitter and similar to dried crab apples. Golden currant is one of the most flavourful and pleasant-tasting currants. Some species, such as stink currant, have a skunky smell and flavour when raw but are delicious cooked.

Sticky currant berries were highly valued by the Haida and other northwest coast First Nations, who collected large quantities to eat fresh with grease or oil (some say this addition was to prevent constipation or stomach cramps). Some First Nations ate wax currants as a strengthening tonic and to treat diarrhea.

In Europe, currant juice is taken as a natural remedy for arthritic pain. Black currant seeds contain gamma-linoleic acid, a fatty acid that has been used in the treatment of migraine headaches, menstrual problems, diabetes, alcoholism, arthritis and eczema.

Some native peoples believed that northern black currant had a calming effect on children, so sprigs were often hung on baby carriers. Currant shrubs growing by lakes were seen as indicators of fish; in some legends, when currants dropped into the water, they were transformed into fish.

Red flowering currant is commonly sold in garden centres in BC as a decorative native shrub, with flowers ranging in colour from pure white to dark red.

The name currant comes from the ancient Greek city of Corinthe, where a small purple grape (*Uva corinthiaca*) is grown and sold commercially as a "currant."

For more information on closely related species, see gooseberries and prickly currants.

Trailing black currant (*R. laxiflorum*)

Golden currant (*R. aureum*)

EDIBILITY: edible

FRUIT: Fruit colour varies from bright red to green to black, as do sweetness and juiciness, depending on the species and individual location. Fruits tart, juicy berries (currants), often speckled with yellow, resinous dots or bristling with stalked glands.

SEASON: Flowers April to July. Fruits ripen July to August.

DESCRIPTION: Erect to ascending, deciduous shrubs, 1–3 m tall, without prickles, but often dotted with yellow, crystalline resin-glands that have a sweet, tomcat odour. Leaves alternate, 3- to 5-lobed, usually rather maple leaf-like. Flowers small (about 5–10 mm across), with 5 petals and 5 sepals, borne in elongating clusters in spring.

Red flowering currant (*R. sanguineum*)

Wax currant (*R. cereum*)

Northern black currant (*R. hudsonianum*)

Red swamp currant (*R. triste*)

Golden currant (*R. aureum*) grows 1–3 m tall and is named for its showy, bright yellow flowers, not its smooth fruits, which range from black to red and sometimes yellow. Its leaves have 3 widely spreading lobes and few or no glands. Inhabits streambanks and wet grasslands to dry prairies and open or wooded slopes in southern BC.

Northern black currant (*R. hudsonianum*) grows to 1.5 m tall, with elongated clusters of 6–12 saucer-shaped, white flowers or shiny, resin-dotted, black berries. Its relatively large, maple leaf–like leaves have resin dots on the lower surface. Fruit strong-smelling and often bitter-tasting. Found in wet woods and on rocky slopes.

Red flowering currant (*R. sanguineum*) is an erect, unarmed, 1–3 m tall shrub with crooked stems and reddish brown bark. Flowers a beautiful and distinctive rose colour (varying from pale pink to deep red), rarely white, 7–10 mm long; in erect to drooping clusters of 10–20 or more flowers. Fruits blue-black, round berries with

glandular hairs and a white, waxy bloom, 7–9 mm long; insipid and seedy. Inhabits dry to moist, open forests and openings at low to middle elevations in southern BC.

Red swamp currant (*R. triste*) is an unarmed, reclining to ascending shrub, to 1.5 m tall. Flowers reddish or greenish purple, small, several (6–15) in drooping clusters; flower stalks jointed, often hairy and glandular. Fruits bright red, smooth and sour but palatable. Found in moist, coniferous forests, swamps, on streambanks and montane, rocky slopes.

Skunk currant (*R. glandulosum*)

Skunk currant (*R. glandulosum*) is a loosely branched, unarmed shrub, 0.5–1 m tall with spreading stems. Bark is brownish. Leaves 3–8 cm wide, 5–7 lobed, glabrous. Flower stalks not jointed below the flowers. Berries dark red, 6–8 mm, nearly round in shape, considered not very nice to eat.

Sticky currant (*R. viscosissimum*) is a loosely branched shrub, to 2 m tall, with erect to spreading stems. Leaves 3- to 5-lobed, heart-shaped at the base, 2–10 cm wide and glandular-sticky. Flowers white or creamy in clusters of up to 16. Fruits bluish black, hairy and glandular berries. Grows in moist to dry forests and woodlands at montane to subalpine elevations in southern BC.

Stink currant (*R. bracteosum*) is a more or less erect, unarmed, straggly, deciduous shrub growing to 3 m tall. All parts of the plant are covered with round, yellow glands that smell sweet-skunky or catty. Leaves alternate, 5–20 cm wide, maple leaf–shaped,

Red swamp currant (*R. triste*)

Trailing black currant (*R. laxiflorum*)

Stink currant (*R. bracteosum*)

deeply 5- to 7-lobed, sweet-smelling when crushed. Flowers white to greenish white; many (20–40) in long, erect clusters 15–30 cm long. Fruits blue-black berries with a whitish bloom, edible, taste variable, from unpleasant to delicious. Found in moist to wet places (woods, streambanks, floodplains, shorelines, thickets, avalanche tracks) at low to subalpine elevations throughout coastal BC.

Trailing black currant (*R. laxiflorum*) is a trailing, spreading plant, occasionally vining, with branches growing along the ground and usually less than 1 m tall. Flowers greenish white to reddish purple in colour. Fruits purplish black, stalked glandular hairs, waxy bloom. Inhabits clearings, disturbed sites such as avalanche

tracks and roadsides and moist forests at mid to low elevations.

Wax currant (*R. cereum*) grows to 1 m tall, with small clusters of 2–6 white to pinkish flowers and red, smooth to glandular-hairy berries. Its shallowly lobed, fan-shaped leaves are glandular on both sides. Found on dry slopes and rocky places in the Interior of BC.

Wax currant (*R. cereum*)

Sticky currant (*R. viscosissimum*)

Sticky currant (*R. viscosissimum*)

Gooseberries   *Ribes* spp.

Northern gooseberry (*R. oxyacanthoides*)

While gooseberries and currants are closely related species, they are generally different in that gooseberries have spines or prickles on their stems (currants are not thus "armed") and gooseberry fruit are usually borne in small clusters or singly (currants are in elongated clusters, generally more than five). However, as common names are inconsistent, some "gooseberries" don't have spines and some "currants" do!

All gooseberries are edible raw, cooked or dried, but flavour and sweetness vary greatly with species, habitat and season. All are high in pectin and make excellent jams and jellies, either alone or mixed with other fruits. Gooseberries can be eaten fresh and are also good in baked goods such as pies. Traditionally, these fruits were eaten with grease or oil, and also mashed (usually in a mixture with other berries) and formed into cakes

Coastal black gooseberry (*R. divaricatum*)

that were dried and stored for winter use. Dried gooseberries were sometimes included in pemmican, and dried gooseberry and bitterroot cakes were sometimes a trade item. Because of their tart flavour, gooseberries can be used much like cranberries. They make a delicious addition to turkey stuffing, muffins and breads. Timing is important, however, when picking these fruit. Green berries are too sour to eat, and ripe fruit soon drops from the branch. Contrary to this, the Nuxalk liked to pick the green berries of coastal black gooseberry and boil them to make a sauce that was much enjoyed. Sometimes green berries can be collected and then stored so that they ripen off the bush. Eating too many gooseberries can cause stomach upset, especially in the uninitiated.

Because of the large number of species and wide distribution of gooseberries, there is a very large spectrum of uses for this genus. They were commonly eaten or used in teas for treating colds and sore throats, a use which may be related to their high vitamin C content. Teas made from gooseberry leaves and fruits were given to women whose uteruses had slipped out of place after too many pregnancies. Gooseberry tea was also used as a wash for soothing skin irritations such as poison-ivy rashes and erysipelas (a condition with localized inflammation and fever caused by a *Streptococcus* infection).

Gooseberries have strong antiseptic properties and extracts have proved effective against yeast (*Candida*) infections.

Picking this fruit can be a formidable task because of the often-thorny stems. Indeed, gooseberry thorns can be so large and strong that they were historically used as needles for probing boils, removing splinters and even applying tattoos! The name "gooseberry" comes from an old English tradition of stuffing a roast goose with the berries.

EDIBILITY: edible

FRUIT: Fruits smooth, purplish (when ripe) berries, about 1 cm across. Dried brown flowerhead is attached at the bottom end of the ripe fruit.

SEASON: Flowers May to June. Fruit ripens July to August.

DESCRIPTION: Erect to sprawling deciduous shrubs with spiny branches. Leaves alternate, maple leaf–like, 3- to 5-lobed, about 2.5–5 cm wide. Flowers whitish to pale greenish yellow, to 1 cm long, tubular, with 5 small, erect petals and 5 larger, spreading sepals, in 1- to 4-flowered inflorescences in leaf axils.

White-stemmed gooseberry (*R. inerme*)

Sticky gooseberry (*R. lobbii*)

Coastal black gooseberry (*R. divaricatum*) is an erect to spreading shrub growing to 2 m tall. Branches arching in form, stems not prickly but have 1–3 stout spines on each node, bark greyish and smooth. Leaves small, maple leaf–shaped, with 3–5 lobes. Flowers white to more commonly red to reddish green, drooping, in clusters of 2–4. Berries round and smooth, purplish black when ripe, with translucent skin. Found in open woods, meadows, moist clearings and hillsides at low elevations on southwestern Vancouver Island and adjacent islands and mainland. Also called: spreading gooseberry.

Mountain gooseberry (*R. irriguum*) is a subspecies of *R. oxyacanthoides* and is differentiated by having a distinctive, long, glandular peduncle and a broad calyx tube with lobes that are longer than the tube.

Coastal black gooseberry (*R. divaricatum*)

91

Northern gooseberry (*R. oxyacanthoides*)

Northern gooseberry (*R. oxyacanthoides*)

Northern gooseberry (*R. oxyacanthoides*) is an erect to sprawling shrub to 1.5 m tall, branches bristly, often with 1–3 spines to 1 cm long at nodes; intermodal bristles absent or few. Twigs grey to straw-coloured, older bark whitish grey. Inhabits wet forests, thickets, clearings, open woods and exposed rocky sites. Also called: smooth gooseberry • *R. hirtellum, R. setosum.*

Northern gooseberry (*R. oxyacanthoides*)

Sticky gooseberry (*R. lobbii*) has clusters of 3 slender spines at each node, is straggly in form and covered with soft, sticky hairs (hence the common name). Bark reddish brown and shredding with age. The fruit of this species is the largest of our native gooseberries (to 1.5 cm diameter) and most resembles the cultivated gooseberry in size and flavour. Found in lowland valleys and on streambanks to open or wooded, montane slopes. Also called: gummy gooseberry.

White-stemmed gooseberry (*R. inerme*) is similar to coastal black gooseberry but has white to pinkish petals (vs. often reddish in coastal black gooseberry), 1–1.5 mm long (vs. 1.5–2.5 mm in coastal black gooseberry). Grows over a different range, in foothill and montane forests in the southern Interior.

Sticky gooseberry (*R. lobbii*)

White-stemmed gooseberry (*R. inerme*)

Coastal black gooseberry (*R. divaricatum*)

Sticky gooseberry (*R. lobbii*)

Prickly Currants   *Ribes* spp.

Prickly currant (*R. lacustre*)

Despite the prickly and strongly irritating stems of these plants, the fruit is quite palatable when ripe and makes delicious jam and pies (either alone or mixed with other fruit) for those who survive the spikes. This is an unusual species in that it has characteristics common to both currants and gooseberries. Like a gooseberry, this species is covered in many sharp, spiny prickles that are also highly irritating. Like a currant, its fruit hangs in clusters.

Although the flavour of bristly black currants is sometimes described as insipid, First Nations people often used them for food. The berries were traditionally eaten fresh off the bush, cooked, stored in the ground for winter use, or sometimes dried. Dried currants were occasionally included in pemmican, and they make a tasty addition to bannock, muffins and breads.

The currants may be boiled to make tea or mashed in water and fermented with sugar to make wine. The leaves, branches and inner bark of prickly currant produce a menthol-flavoured tea, sometimes called "catnip tea." To make this tea, the spines were singed off, and the branches (fresh or dried) were steeped in hot water or boiled for a few minutes. Bristly currant fruits were traditionally rolled on hot ashes to singe off their soft spines before eating. Both red currant jelly and the juice of green currants are said to be correctives for spoiled food (especially high, or aged and slightly decomposed, meat), but this use is not recommended.

The hairy appearance of *R. lacustre* berries prompted some BC First Nations to give this fruit a common name of "hairy face" berry. Some native peoples considered the spiny branches (and by extension, the fruit) to be poisonous. However, these dangerous shrubs could also be useful as their thorny branches were thought to ward off evil forces.

Prickly currant (*R. lacustre*)

Prickly currant
(*R. lacustre*)

EDIBILITY: edible

FRUIT: Shiny black berry, hanging in drooping clusters and covered in long hairs.

SEASON: Blooms in April. Fruits ripen May to August.

DESCRIPTION: Erect to spreading deciduous shrubs with spiny, prickly branches and alternate, 3- to 5-lobed, maple-like leaves. Flowers reddish to maroon, saucer-shaped, about 6 mm wide, forming hanging clusters of 7–15. Fruits 5–8 mm-wide berries, bristly with glandular hairs.

Mountain prickly currant (*R. montigenum*)

Prickly currant (*R. lacustre*)

Prickly currant (*R. lacustre*)

Prickly currant (*R. lacustre*)

Mountain prickly currant (*R. montigenum*) is similar to bristly black currant but has smaller leaves (generally 1–2.5 cm wide) that are glandular-hairy on both sides, and bright red berries. Grows on rocky montane, subalpine and alpine slopes in southwestern BC. Also called: alpine prickly currant.

Prickly currant (*R. lacustre*) is an erect to spreading shrub, 50 cm to 2 m tall, covered with numerous small, sharp prickles, and larger, thick thorns at leaf nodes. Bark on older stems is cinnamon coloured. Leaves large (2–7 cm wide) and hairless to slightly hairy. Berries dark purple. Found in moist woods and streambanks to drier forested slopes and subalpine ridges. Also called: swamp gooseberry, bristly black currant, swamp black currant.

Prickly currant (*R. lacustre*)

Saskatoon   *Amelanchier alnifolia*

Also called: serviceberry, Canada serviceberry, juneberry, shadbush • *A. florida*

Saskatoon (*A. alnifolia*)

These sweet fruits were and still are extremely important to many indigenous peoples across Canada. Indeed, there is a well-documented history of extensive landscape management through techniques such as fire, weeding and pruning to encourage the healthy growth of this important species. Large quantities of the berries were harvested and stored for consumption during winter. They were eaten fresh, dried like raisins, or mashed and dried into cakes for winter use or trade. Some indigenous peoples steamed saskatoons in spruce bark vats filled with alternating layers of red-hot stones and fruit. The cooked fruit was mashed, formed into cakes, then dried over a slow fire. These cakes could weigh as much as 7 kg (15 lbs) each!

Dried saskatoons were the principal berries mixed with meat and fat to make pemmican and were commonly added to soups and stews. Today, they are popular in pies, pancakes, puddings, muffins, jams, jellies, sauces, syrups and wine, much like blueberries.

Historically, saskatoon juice was taken to relieve stomach upset and was also boiled to make drops to treat earache while green or dried berries were used to make eye drops. The fruits were given to mothers after childbirth for afterpains and were also prescribed as a blood remedy. The berry juice, which can stain your hands when picking, makes a good purple-coloured dye.

Saskatoons are excellent ornamental, culinary and wildlife shrubs. They are hardy and easily propagated, with beautiful white blossoms in spring, delicious fruit in summer and colourful, often scarlet leaves in autumn. There are a number of improved garden cultivars available for superior fruit production in the home garden.

EDIBILITY: highly edible

FRUIT: Fruits juicy, berry-like pomes, red at first, ripening to purple or black, sometimes with a whitish bloom, 6–12 mm across.

SEASON: Flowers April to June. Fruit ripens July to August.

DESCRIPTION: Shrub or small tree, to 7 m tall, often forming thickets. Bark is smooth, grey to reddish brown. Leaves alternate, coarsely toothed on the upper half, leaf blades 2–5 cm long, oval to nearly round, yellowish orange to reddish brown in autumn. Flowers are white, forming short, leafy clusters near the branch tips; petals are slightly twisted. Grows at low to middle elevations in prairies, thickets, hillsides and dry, rocky shorelines, meadows, open woods throughout BC, particularly in the Interior.

WARNING: *The fresh leaves and pits contain poisonous cyanide-like compounds. However, cooking or drying destroys these toxins.*

Saskatoon Squares

¼ cup butter • ⅔ cup brown sugar
1 tbsp vanilla • 1 large egg, beaten
1 cup flour • 1 tsp baking powder
½ tsp salt • ½ tsp cinnamon
½ cup FROZEN saskatoons (or wild blueberries)
½ cup chopped walnuts or almonds

Preheat oven to 350° F. Melt the butter gently in a saucepan, then remove from heat and stir in sugar, vanilla and beaten egg. Mix dry ingredients in a bowl. Make a shallow well in the middle, and gradually mix in wet ingredients from the saucepan. When it's well mixed, add frozen saskatoons and chopped nuts. Pour into an 8-inch pan and bake for 35 minutes. Remove from oven and cool before cutting into squares.

Pemmican

Makes 6 cups

The pemmican uses the same drying temperature as the fruit leather (see p. 73), so make both recipes at the same time!

3 Tbsp salted butter • 3 Tbsp brown sugar
¼ tsp dried powdered ginger
¼ tsp ground cloves • ¼ tsp ground cinnamon
4 cups saskatoons or blueberries
4 cups beef jerky, chopped into small pieces
½ cup chopped almonds, walnuts or hazelnuts (optional)
½ cup sunflower seeds (optional)

Gently heat butter with sugar and spices in a heavy-bottomed pot. Mash berries and add to pot. Simmer, stirring constantly, for about 5 minutes. Let mixture cool, then mix in jerky and nuts and/or seeds. Grease a rimmed baking sheet, spread mixture evenly on sheet and let dry overnight in oven at 150° F.

Red-osier dogwood (*C. sericea*)

The fruits of the Pacific dogwood tree and its shrubbier cousin the red-osier dogwood are definitely bitter to modern-day tastes. While Pacific dogwood produces fruit of an attractive red colour that look like they should be delicious, the taste is hard, mealy and bitter. The berries of red-osier dogwood, however, despite their bitterness, were gathered by some First Nations in late summer and autumn and eaten immediately. They were also occasionally stored for winter use, either alone or mashed with sweeter fruits such as saskatoons, and in more modern times with sugar. The Ktunaxa people of the southern BC Interior make a dish called "sweet and sour," which is a mixture of red-osier dogwood fruit, a sweet berry such as saskatoon and a bit of sugar. The red-osier dogwood fruit can be cooked when fresh to release the juice, which purportedly makes a refreshing drink when sweetened. Some people separated the stones from the mashed flesh and saved them for later use. They were then eaten as a snack, somewhat like peanuts are today. However, this is not recommended in large quantities and the taste is probably not worth the effort involved. Pacific dogwood and red-osier dogwood are both popular and attractive ornamental trees with good wildlife and aesthetic values.

Pacific dogwood (*C. nuttallii*)

Pacific dogwood (*C. nuttallii*)

Pacific dogwood (*C. nuttallii*)

EDIBILITY: edible with caution (toxic)

FRUIT: Fruits of Pacific dogwood are bright red, berry-like drupes in clusters of at least 20, with each drupe small, but growing up to 1 cm in size. Although pretty to look at, they are mealy and hard in consistency, often bitter tasting, and not considered edible—perhaps best left to the birds or as a famine food if you get lost in the woods. Fruits of red-osier dogwood are white (sometimes bluish), pea-sized, berry-like drupes, to 9 mm across, containing large, flattened stones. It is reported that the whiter fruits, though bitter, are less so than the bluer-tinged ones.

SEASON: Flowers May to June. Berries ripen late summer, are edible when coloured red (or white/bluish for red-osier dogwood), and are tradition-ally gathered from August to October.

DESCRIPTION: Erect to sprawling, deciduous shrubs or small trees with opposite branches. Leaves opposite, simple, pointed, toothless, with leaf veins following the smooth leaf edges toward the tips, greenish above, white to greyish below, becoming red in autumn. Tiny inconspicuous flowers surrounded by petal-like bracts that resemble a single flower. Flowers radially symmetrical with 4 sepals, petals and stamens, all attached at the top of the ovary. Fruit a fleshy, berry-like drupe.

Pacific dogwood (*C. nuttallii*) grows to 20 m tall, with blackish-brown, smooth bark, becoming finely ridged with age. Branches opposite, greyish-purplish when young. Flowers clusters of small, yellowish green or red flowers, about 75 in a cluster to 2 cm wide, subtended by 4–7 white- or pinkish-tinged, showy bracts. Note that to the common observer these showy bracts appear as large "petals" of a giant flower, while the tiny true flowers appear as a decorative centre to that main display. Grows in moist, well-drained soils in the shade of coniferous trees in southwestern BC and inhabits a wide range of habitat from moist valley bottoms to just below higher elevation timberline. Pacific dogwood is the provincial flower of BC and as such it is against the law to harvest or cut down the tree. Also called: western flowering dogwood.

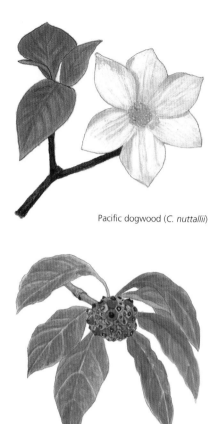

Pacific dogwood (*C. nuttallii*)

Pacific dogwood (*C. nuttallii*)

Pacific dogwood (*C. nuttallii*)

Red-osier dogwood (*C. sericea*). Two varieties of this species grow in BC: *C. sericea* var. *occidentalis* grows west of the Cascades while *C. sericea* var. *stolonifera* (differentiated mainly by having smooth stones in the fruit rather than rough) grows east of this mountain range. Erect to sprawling, deciduous shrub or small tree, slender and branching in form, to 6 m tall. Twigs and branches are opposite, shiny and smooth, bright green to red when young (the more sun exposure on the stem, the brighter the red colour), changing to brown when older. Leaves opposite, simple, pointed, toothless, with leaf veins following the smooth leaf edges toward the tips, greenish above, white to greyish below, becoming red in autumn. Flowers small, clustered, 2–4 cm wide, without showy bracts, from May to August. Fruits white (sometimes bluish), pea-sized, berry-like drupes, to 9 mm across, containing large, flattened stones. Grows on moist sites, shores and thickets throughout Canada. Also called: western dogwood, red-willow • *C. alba, Svida sericea*.

Red-osier dogwood (*C. sericea*)

Red-osier dogwood (*C. sericea*)

Red-osier dogwood (*C. sericea*)

Red-osier dogwood (*C. sericea*)

WARNING: *All parts of this species are considered toxic, especially if consumed in large quantities.*

Huckleberries *Vaccinium* spp.

Velvetleaf huckleberry (*V. myrtilloides*)

Huckleberries are delicious and well worth identifying and eating. These berries are generally sweet but tend to be tart and strong-flavoured (blueberries, on the other hand, tend to be purely sweet and cranberries sharply tart). Huckleberry fruit tend to be blackish and glossy while blueberries are generally blue in colour with a whitish bloom on the fruit.

Huckleberries can be used like domestic blueberries, eaten fresh from the bush, added to fruit salads, cooked in pies and cobblers, made into jams and jellies or crushed and used in cold drinks. They are also delicious in pancakes, muffins, cakes and puddings. Dried huckleberry leaves and berries also make excellent teas.

Black huckleberry (*V. membranaceum*)

Grouseberry (*V. scoparium*)

Evergreen huckleberry (*V. ovatum*)

On the West Coast, many First Nations considered evergreen huckleberry to be one of the tastiest berries and traditionally travelled considerable distances to collect it. These fruit were eaten fresh or dried for winter use and trade. While they ripen in early autumn, evergreen huckleberry fruit can remain on the plants until December and some reports claim that they taste even better after a frost.

Some people consider black huckleberry the most delicious and highly prized berry in western Canada and the Rockies. Black huckleberries are collected in large quantities even today in open, subalpine sites (such as old burns), and in some regions they are sold commercially.

Bears and other wildlife relish huckleberries.

Huckleberries were considered good for the liver by some First Nations and were eaten as a ceremonial food to ensure health and prosperity for the coming season. Red huckleberries were traditionally eaten fresh, dried or made into cakes, or mixed with grease or oil to preserve them for winter use, gifts or as a trade item. The berries were mashed and formed into cakes or spread loosely on mats for drying. Later, they were reconstituted by boiling either alone or with other foods.

Evergreen huckleberry (*V. ovatum*)

Red huckleberry fruit is tart and delicious and in a good patch it is easy to harvest it in large quantities. This berry tastes similar to a cranberry, but is milder and sweeter and is excellent eaten fresh. If you have enough willpower to gather some in a bucket to take home rather than eating them all in situ, these berries make particularly good jam and pies. You can use red huckleberries wherever cranberries are called for in a recipe, both sweet and savoury. Red huckleberry makes an excellent addition to the home garden, doing well in a moist, shady location that many other plants reject. It also does well in moist areas that have full sun. An established red huckleberry specimen will reward the homeowner with a pretty display of delicate green leaves in the spring and a delicious crop of decorative bright berries in late summer. It will attract wildlife (competition for said berries!), and at the end of the year provide

a striking winter display with pinkish coloured stems that look lovely in winter floral arrangements. More recently, the winter branches of red huckleberry have become a popular greenery in the florist trade.

The berries of grouseberry are sweet and wonderful, but their small size makes picking them rather time-consuming. Try using a (clean) comb to rake the berries into a basket, hat, or other container to speed up this process. They can grow quite abundantly in alpine areas, so a little persistence can yield a good haul. Try to get there early, though—this berry is a favourite food of chipmunks, red and grey foxes, squirrels, skunks and many bird species such as blue, spruce and ruffed grouse as well as ptarmigan, bluebirds and thrushes. The plants themselves are a favourite browse for mule deer, bear, moose and mountain goat. Tough competition indeed!

Black huckleberry
(*V. membranaceum*)

Red huckleberry (*V. parvifolium*)

EDIBILITY: highly edible

FRUIT: Berries range in colour from red to purple to black.

SEASON: Flowers bloom April to June. Berries ripen July to September.

DESCRIPTION: Deciduous or evergreen shrubs, 40 cm to 3 m tall, with alternate, 2–5 cm-long leaves. Flowers various shades of pink, round to urn-shaped, 4–6 mm long, nodding on single, slender stalks, from April to June. Fruits are berries, 6–10 mm across. All *Vaccinium* species have a small "crown" at the bottom end of the berry, which is a residual left over from the flower that was pollinated to form the fruit.

Vaccinium **spp.**

Common names can be confusing, especially with the large variety of Vaccinium *species we have here in BC. Blueberries, cranberries and huckleberries are all closely related plants in this plant family. In North America there are approximately 35 different* Vaccinium *species, but hybridization is common in the genus so the true number of varieties is probably greater. As a general rule, species of* Vaccinium *with blue fruits are called blueberries, and taller shrubs with fruits that aren't blue are called huckleberries. Shorter* Vaccinium *species with red berries and a distinctive tart flavour are commonly referred to as cranberries. However, common names do not necessarily follow this botanical protocol. For example, high bush cranberry (*Viburnum edule*) is in the honeysuckle family and is not a "true" cranberry at all, despite its red colour and tart flavour. Rest assured, though, none of these species are poisonous and they are all delicious to eat!*

Grouseberry (*V. scoparium*)

Black huckleberry (*V. membranaceum*) is a deciduous shrub with purplish-black berries. Branches greenish when young but greyish brown with age and at most slightly angled. Grows in dry to moist forests at montane to subalpine elevations on open or wooded slopes. Also called: thinleaf huckleberry, black mountain huckleberry.

Evergreen huckleberry (*V. ovatum*) is an erect, bushy shrub to 4 m. Branches greenish when young but brown with age and at most slightly angled. Leaves evergreen, shiny, leathery, paler on the lower surface, egg-shaped, 2–5 cm. Flowers clustered in 3–10, to 8 mm long. Berries shiny, deep purplish black, 4–7 mm. Grows at low to montane elevations in dry to moist forests in coastal BC. Also called: California huckleberry.

Black huckleberry
(*V. membranaceum*)

Evergreen huckleberry
(*V. ovatum*)

Grouseberry (*V. scoparium*)

Grouseberry (*V. scoparium*) is a low, creeping, broom-like perennial shrub to 25 cm tall. Branches are many, squarish, slender, pale green, strongly angled and grow so closely together that they can be bundled together and used as a broom once the leaves have fallen in autumn. Leaves thin, finely toothed, deciduous, alternate, ovate, 6–15 mm long. Flowers tiny, 3–4 mm long, 4 petals fused into urn-shaped blooms, pink, solitary. Berries single, bright red, small 3–5 mm in diameter, smelling tantalizingly like huckleberry jam when ripe! Inhabits open coniferous forest, foothill, montane and subalpine slopes from medium to high elevation in BC, particularly the Rockies in lodgepole pine areas. Also called: grouse whortleberry, littleleaf huckleberry.

Red huckleberry (*V. parvifolium*) is a deciduous shrub to 4 m tall. Branches green and strongly angled, often pinky coloured in winter. Leaves light green, alternate, oval, to 3 mm long, often remaining on the plant through mild coastal winters. Flowers bell to urn-shaped, greenish white to pinkish yellow, 5 mm long. Berries up to 1 cm in diameter, bright pink to reddish orange or red. Berries do not all ripen at once. Grows in dry to moist coniferous forests and forest edges in low to medium elevations and montane zones in BC, most abundantly on the coast.

Velvetleaf huckleberry (*V. myrtilloides*) is a low shrub, to 40 cm tall, with densely hairy (velvety) branches, especially when young. Produces dark bluish-black to dark red fruits, from August to October. Grows at montane elevations in dry to moist forests and openings and bogs.

Velvetleaf huckleberry (*V. myrtilloides*)

Red huckleberry (*V. parvifolium*)

Red huckleberry (*V. parvifolium*)

LITERARY REFERENCES: *The small, dark, rather insignificant fruit of the huckleberry became a metaphor in the early part of the 19th century for something humble or minor. This later came to mean somebody inconsequential, an idea that was the basis for Mark Twain's famous character, Huckleberry Finn. In an 1895 interview, Twain revealed that he wanted to establish the boy as "someone of lower extraction or degree, than Tom Sawyer." The modern word "huckleberry" is an American distortion of the British "whortleberry," a name that originated from the Anglo-Saxon word* wyrtil, *meaning "little shrub" and* beri *meaning "berry."*

False huckleberry (*Menziesia ferruginea*)

False huckleberry (*Menziesia ferruginea*) is a **poisonous** plant that has small urn-shaped flowers that look very similar to *Vaccinium* species. The fruit of this plant is a small many-seeded capsule. However, sometimes there are small pink berry-like "fruit" on the underside of its leaves—this "berry" is actually the fruiting body of a fungus (*Exobasidium vaccinii*). While all parts of this plant are poisonous, these fungal "berries" are apparently edible. According to Tsimshian legend, this fungus (which they ate) is the snot of Henaaksiala, who was a mythical creature who stole corpses.

> **WARNING:** *The plant false huckleberry* (Menziesia ferruginea) *is poisonous, as are its fruit. The edible "berry" sometimes found on the underside of its leaves is the fruiting part of a parasitic fungus.*

False huckleberry (*Menziesia ferruginea*)

Velvetleaf huckleberry (*V. myrtilloides*)

Tom's Huckleberry Pie

Makes 1 double-crust pie

The secret to a good, crispy pastry that is not tough and "dough-like" is to keep all your ingredients cool. Warm ingredients melt the small fat globules, causing them to mix too completely with the flour and resulting in chewy pastry. Leftover pastry trimmings make excellent little tartlets if rolled out again and put into the bottom of muffin tins, and filled with any extra huckleberry filling or jam out of the fridge.

Pastry
2 cups all-purpose flour · ½ tsp salt
⅔ cup vegetable shortening · ⅓ cup COLD milk

Filling
3 cups red or black huckleberries · ¼ cup water or freshly squeezed orange juice
1 cup granulated white sugar · 3 Tbsp cornstarch

For the pastry, sift the flour and salt together, then use 2 butter knives or a pastry cutter to cut the shortening into the flour mixture until the butter pieces are the size of small peas. *Avoid using your hands at this stage because their warmth will cause the butter to melt.* Gradually stir in the cold milk, then quickly shape the dough into 2 balls. Wrap them in plastic film, press into flat rounds and refrigerate immediately.

For the pie filling, mash the huckleberries with the water and put into a medium saucepan. Mix the cornstarch and sugar together and stir well into the COLD berries and water. *Do not heat the berries and water first because the cornstarch will cook prematurely and go all nasty and lumpy!* Bring mixture slowly to the boil, stirring constantly to avoid the cornstarch sticking or getting lumps. Simmer until noticeably thick, about 4 to 5 minutes, then take the saucepan off the burner.

Preheat oven to 350° F. Take the pastry out of the fridge, spread a thin layer of flour on a work surface, and roll the pastry until it is approximately ¼ inch thick. Place it into an 8-inch pie tin, cutting any extra off from around the edges. Then roll out the second half of your dough into a similar-sized round. Fill the pastry shell with the huckleberry mixture, then carefully place the second round on top. Gently push the edges of the top and bottom pastry crusts together (you may need to lightly wet one edge to get it to stay together), and prick a few holes with a fork in the top to allow steam to escape during cooking. Bake for approximately 50 minutes.

Grouseberry (*V. scoparium*)

Blueberries *Vaccinium* spp.

Alaska blueberry (*V. alaskaense*)

Blueberry fruit tend to be blue in colour, hence the common name of this group of plants. The fruit of this species is generally sweet, rather than tart/sour (cranberries) or sweet/tangy (huckleberries). Wild blueberries are simply delicious! Rich in vitamin C and natural antioxidants, these fruit are both beautiful to look at and good for you to eat. The sweet, juicy berries can be eaten fresh from the bush, added to fruit salads, cooked in pies, tarts and cobblers, made into jams, syrups and jellies, or crushed and used to make juice, wine, tea and cold drinks. Blueberries also make a prime addition to pancakes, muffins, cakes and puddings as well as to savoury treats like chutneys and marinades.

These wild fruits were widely used by First Nations, either fresh, dried singly or mashed and formed into cakes. To make dried cakes, the berries were cooked to a mush to release the juice, spread into slabs, and dried on a rack (made from wood, rocks, or plant materials) in the sun or near a fire. Often, the leftover juice was slowly poured onto the drying cakes to increase their flavour and sweetness.

Because blueberries grow close to the ground, they can be difficult and time-consuming to collect, so some people developed a method of combing them from the branches with a salmon back-bone or wooden comb or rake. While this method is efficient, it results in baskets full of both berries and small hard-to-pick-out blueberry leaves. The savvy solution developed for this problem was to place a wooden board at a medium angle and slowly pour the berry/leaf mix from the top of the board: the berries (which are round) roll down the board to a basket waiting below, but the leaves (which are flat) stick to the board and stay put. After two to three rollings, the picker ends up with a basket of pure berries at significantly less effort than would have been required to pick the leaves out individually. The only drawback to this method is that the occasional green berry also gets picked, but these are easily removed by hand.

While most people only associate blueberries with a delicious fruit, there are many historical medicinal uses for other parts of this wide-ranging plant.

Blueberry roots were boiled to make medicinal teas that were taken to relieve diarrhea, gargled to soothe sore mouths and throats, or applied to slow-healing sores. Bruised roots and berries were steeped in gin, which was to be taken freely (as much as the stomach and head could tolerate!) to stimulate urination and relieve kidney stones and water retention. Blueberry leaf tea and dried blueberries have historically been used like cranberries to treat diarrhea and urinary tract infections.

Bog blueberry (*V. uliginosum*)

Oval-leaved blueberry (*V. ovalifolium*)

Dwarf blueberry (*V. caespitosum*)

Dwarf bilberry contains anthocyano-sides, which are said to improve night vision. These compounds are most concentrated in the dried fruit, preserves, jams and jellies. Their effect, however, is said to wear off after five to six hours. Anthocyanins may reduce leakage in small blood vessels (capillaries), and blueberries have been suggested as a safe and effective treatment for water retention during pregnancy, for hemorrhoids, varicose veins and similar problems. They have also been recommended to reduce inflammation from acne and other skin problems and to prevent cataracts. Blueberry leaf tea has been used by people suffering from hypoglycemia and by some diabetics, to stabilize and reduce blood sugar levels, and to reduce the need for insulin. The leaf or root tea of dwarf bilberry is reported to flush pinworms from the body. The leaves of blueberry were sometimes dried and smoked and the berries have been used to dye clothing a navy blue colour.

EDIBILITY: highly edible

FRUIT: Berries round, 5–8 mm wide, bluish coloured, growing in clusters, usually with a greyish bloom.

SEASON: Blooms May to July. Fruit ripen July to September.

DESCRIPTION: Low, often matted shrubs with thin, oval leaves 1–3 cm long. Flowers whitish to pink, nodding, urn-shaped, 4–6 mm long.

Alaska blueberry (*V. alaskaense*) is very similar to oval-leaved blueberry, but has flowers appearing before the leaves (vs. with the leaves in the oval-leaved), flowers that are usually wider than long (vs. longer than wide in oval-leaved), and leaves with coarse, stiff hairs along the underside midrib (vs. bare in oval-leaved). Flowers from April to May. Grows at low to subalpine elevations in mesic to moist forests and openings in BC, mainly coastal.

Alaska blueberry (*V. alaskaense*)

Bog blueberry (*V. uliginosum*)

Dwarf blueberry (*V. caespitosum*)

Oval-leaved blueberry (*V. ovalifolium*)

Bog blueberry (*V. uliginosum*) is a low, spreading, deciduous, perennial shrub, 10–60 cm tall (but can grow as short as 2.5 cm in areas of heavy snow where the shrub is crushed flat each winter). Branches are rounded, brownish. Leaves 3 cm long, alternate, fuzzy to smooth, elliptical to oval, narrow and broadest toward the tip, dull/whitish green, toothless. Flowers solitary or paired, 4–5 lobed, white or pinkish, urn shaped, up to 6 mm long. Berries dark blue to blackish with a whitish bloom, to 9 mm in size. Inhabits low elevation bogs, boggy forests, subalpine heath, and alpine slopes/tundra in the northern Interior of the province. Also called: bog bilberry, western huckleberry • *V. occidentale*.

Dwarf blueberry (*V. caespitosum*) is a low, usually matted shrub, 10–30 cm tall, with rounded, yellowish to reddish branches and finely toothed, light green leaves. Its 5-lobed flowers produce berries singly in leaf axils, from August to September. Grows at all elevations in dry to wet forests, bogs, meadows, rocky ridges and tundra throughout Canada.

Oval-leaved blueberry (*V. ovalifolium*) is a tall shrub, to 2 m tall, with hairless, angled branches. Its leaves are entire or only sparsely toothed. Produces purple berries with a whitish bloom from early July to September. Grows at low to subalpine elevations in dry to moist forests, openings and bogs.

Cascade bilberry (*V. deliciosum*) is a low, often matted and densely-branched shrub 15–40 cm tall. Stems greenish brown, minutely hairy or glabrous, inconspicuously angled. Leaves to 3 cm long, pale green with a whitish bloom beneath, often minutely rolled downward. Fruit 6–8 mm, almost globe-shaped, singly or in pairs, blue to blue-black, ripening early July to September. Inhabits dry to moist forests, open areas and bogs at low to subalpine elevations.

Dwarf bilberry (*V. myrtillus*) is a low, many-branched shrub to 30 cm tall. Stems strongly angled, greenish brown, minutely hairy. Leaves light green, 1–3 cm long, elliptic-lanceolate to egg-shaped, alternate, sharply toothed margins, prominent rib on lower surfaces, strongly veined. Flowers 2–3 mm, pinkish, in leaf axils, solitary, on nodding stalks. Blooms May to July. Fruit ripen July to October. Fruit 4–8 mm, globe shaped, dark red to bluish black in colour without a whitish bloom. Easy to confuse with *V. scoparium*, but *V. myrtillus* has a more open habit with minutely hairy branches that are thicker and less numerous (*V. scoparium* is more broom-like) as well as being all-around slightly larger and with darker-coloured fruits. Bilberries are similar in appearance and taste to blueberries, but bilberry fruit grow singly or in pairs while blueberry fruit grow in clusters. Note that the common name "whortleberry" also refers to *Arctostaphylos alpina*. Inhabits dry mesic (modest to well-balanced moisture) forests, wooded montane and subalpine slopes in the Kootenay and Rockies areas of BC. Also called: low bilberry, whortleberry • *V. oreophilum.*

Dwarf bilberry (*V. myrtillus*)

WARNING: *Blueberry leaves contain moderately high concentrations of tannins, so they should not be used continually for extended periods of time.*

Dwarf bilberry (*V. myrtillus*)

Oval-leaved blueberry (*V. ovalifolium*)

Dwarf blueberry (*V. caespitosum*)

Blueberry Cobbler

1 cup flour • 2 Tbsp sugar • 1½ tsp baking powder
¼ tsp salt • 1 tsp grated lemon zest • ¼ cup butter
1 beaten egg • ¼ cup milk • 2 Tbsp corn starch
½ cup sugar • 4 cups fresh blueberries (or huckleberries)

Preheat oven to 425° F. Sift together all the dry ingredients. Mix wet ingredients together, then pour slowly into the dry mix, stirring until just moistened.

Mix cornstarch and sugar together, and toss with the fruit. Pour this mixture into the bottom of an 8 x 10-inch glass or ceramic baking dish (avoid metal dishes because the acid in the fruit might turn it rusty and impart a nasty flavour to the cobbler). Drop the topping in spoonfuls on top of the fruit, covering the surface as evenly as possible (some exposed areas of fruit are fine). Bake, uncovered, for 25 minutes or until light brown.

Fruit Popsicles

Makes 8 to 12 popsicles

Easy and a popular treat for adults and kids at any time of the year!

4 cups wild berries • 1 cup plain yogurt or light cream
1 cup white sugar • 1 cup orange or other fruit juice (optional)

Blend all ingredients together, pour into the compartments of a popsicle maker and place in freezer until frozen.

Cranberries   *Vaccinium* spp.

Bog cranberry (*V. oxycoccos*)

Like many other species of wild berries with domesticated counterparts, wild cranberries are small but packed with a flavour that seems disproportionate to their size. The tartness of cranberries gives them an enviable versatility for sweet, sour and savoury dishes. Who could imagine Thanksgiving dinner or many juices, desserts or baking without them? These berries, which also have a long history of medicinal use in treating kidney and urinary ailments, are touted today for their strong antioxidant properties. (Note: high bush cranberries, while tart and cranberry-tasting, are in the honeysuckle family and are therefore treated in a separate account.)

Cranberries can be very tart, but they make a refreshing trail nibble. They make delicious jams and jellies, and they can be crushed or chopped to make tea, juice or wine. A refreshing drink is easily made by simmering berries (crush them first to allow the juice out more easily) with sugar and water or, more traditionally, by mixing cranberries with maple sugar and cider. Cranberry sauce is still a favourite with meat or fowl and the berries add a pleasant zing to fruit salads, pies and mixed fruit cobblers. Cranberries are also a delicious addition to pancakes, muffins, breads, cakes and puddings.

Firm, washed berries keep for several months when stored in a cool place. They can also be frozen, dried or canned. First Nations sometimes dried cranberries for use in pemmican, soups, sauces and stews. Some tribes stored boiled cranberries mixed with oil and later whipped this mixture with snow and fish oil to make a dessert.

Freezing makes cranberries sweeter, so they were traditionally collected after the first heavy frost. Because they remain on the shrubs all year, cranberries can be a valuable survival food rich in vitamin C and antioxidants. These low-growing berries are difficult to collect, so some people combed them from their branches with a salmon backbone or wooden comb.

Cranberry juice has long been used to treat bladder infections. Research shows that these berries contain arbutin, which prevents some bacteria from adhering to the walls of the bladder and urinary tract and causing an infection. Cranberry juice also increases the acidity of the urine, thereby inhibiting bacterial activity, which can relieve infections. Commercial cranberry juice or cocktail blends are not appropriate for this treatment, however, as the juice is highly processed, often diluted with other juices and is highly sweetened (sugars will feed the problem bacteria and can make the existing condition worse). Increased acidity can also lessen the urinary odour of people suffering from incontinence. The tannins in cranberry have anti-clotting properties and are able to reduce the amount of dental plaque-causing bacteria in the mouth, thus are helpful against gingivitis. Research has shown that cranberries contain antioxidant polyphenols that may be beneficial in maintaining cardiovascular health and immune function, and in preventing tumour formation. Although some of these compounds have proved extraordinarily powerful in killing certain types of human cancer cells in the laboratory, their effectiveness when ingested is unknown. There is also evidence that cranberry juice may be effective against the formation of kidney stones.

Cranberries were traditionally prescribed to relieve nausea, to ease cramps in childbirth and to quiet hysteria and convulsions. Crushed cranberries were used as poultices on wounds, including poison-arrow wounds. The red pulp

Lingonberry (*V. vitis-idaea*)

(left after the berries have been crushed to make juice) can be used to make a red dye. A related species, lingonberry, is popular in Sweden as a digestive aid, in jams, jellies, pies and other baking, juice, wine liqueur, herbal tea, and as "cranberry sauce." The Inupiat Inuit used a wrapped cloth containing mashed berries to treat sore throats and crushed berries to treat itchy skin conditions such as chicken pox or measles. The mashed fruit, mixed with oil, was fed to convalescents to help them recover strength.

Bog cranberry (*V. oxycoccos*)

Bog cranberry (*V. oxycoccos*)

EDIBILITY: highly edible

FRUIT: Berries bright red (sometimes purplish) when ripe, 6–10 mm wide. Tart and delicious, said to be best after the first frost, and can be foraged when snows melt in the spring—if the wildlife have left any behind!

SEASON: Flowers June to July. Fruit ripens August to September and may persist on plants throughout the winter.

Bog cranberry (*V. oxycoccos*)

Bog cranberry (*V. oxycoccos*)

Cranberry Chicken

Serves 5

3 lbs chicken • ¼ cup flour • ½ tsp salt
¼ cup cooking oil • 1½ cups fresh cranberries
½ cup sugar • 1 Tbsp grated orange zest
½ cup fresh orange juice • ¼ tsp ground ginger

Cut chicken into serving-sized pieces, and coat
with flour and salt. Heat oil in a cast-iron skillet.
Add chicken pieces and brown on both sides,
being careful to cook the chicken fully. Combine
remaining ingredients in a saucepan, bring to
the boil and pour it over the chicken in the
skillet. Cover skillet, reduce heat and simmer
30 to 40 minutes until chicken is tender.

DESCRIPTION: Dwarf, low-spreading
deciduous shrubs, mostly less than
20 cm tall, often trailing, with small,
nodding, pinkish flowers producing
sour, bright red (sometimes purplish)
cranberries.

Bog cranberry (*V. oxycoccos*) has
slender, creeping stems, with small
(mostly less than 1 cm long), pointed,
glossy leaves. Flowers are distinctive,
4 petals separated almost to the base,
the petals curved strongly backward
(like little shooting stars), flowers
appearing terminal on stems. Fruit
a deep red, globose cranberry, about
5–12 mm wide, from July to August.
Bog cranberry predictably inhabits
bogs. Also called: small cranberry
• *Oxycoccus oxycoccos, O. quadripetalus,
O. microcarpus.*

Lingonberry (*V. vitis-idaea*) is
a low-spreading shrub with rounded
or slightly angled branches. Leaves
evergreen, 6–15 mm long, blunt,
leathery, with dark dots (hairs) on
a pale lower surface. Flowers 4-petalled,
fused into small urn-shaped nodding

Lingonberry (*V. vitis-idaea*)

blooms, pinkish in colour, 1 to several
in terminal clusters. Fruit bright red
cranberries, 6–10 mm wide, from July
to August. Grows in acidic soils in
sunny mountain meadows, peat moors,
dry woods, foothill, montane, subalpine
and alpine slopes. Also called: mountain
cranberry, low bush cranberry, rock
cranberry, cowberry, partridge berry.

False-wintergreens   *Gaultheria* spp.

Slender false-wintergreen (*G. ovatifolia*)

The small, sweet berries of this species are delicious and can be eaten fresh, served with cream and sugar or cooked in sauces. Their flavour improves after freezing, so they are at their best in winter following the first frost (even from under the snow if you are persistent!), or in spring when they are plump and juicy. The young leaves can be an interesting trailside nibble or added to salads as well as used to make a strong, aromatic tea that is said to make a good digestive tonic. The wintergreen flavour can be drawn out if the bright red leaves are first fermented.

The berries were historically mixed with teas and were used to add fragrance and flavour to liqueurs. Occasionally, large quantities were picked and dried like raisins for winter use. During the American Revolutionary War, wintergreen tea was a substitute for black tea (*Camellia sinensis*). The berries were traditionally soaked in brandy, and the resulting extract was taken to stimulate appetite, as a substitute for bitters.

All false-wintergreens contain methyl salicylate, a close relative of aspirin that has been used to relieve aches and pains. These plants were widely used in the treatment of painful, inflamed joints resulting from rheumatism and arthritis (see Warning). Studies suggest that oil of wintergreen is an effective painkiller and it also has numerous commercial applications. It provides fragrance to various products such as toothpastes, chewing gum and candy and is used to mask the odours of some organophosphate pesticides. It is a flavouring agent (at no more than 0.04 percent) and an ingredient in deep-heating sports creams. The oil is also a source of triboluminescence, a phenomenon in which a substance produces light when rubbed, scratched or crushed. The oil, mixed with sugar and dried, builds up an electrical charge that releases sparks when ground, producing the Wint-O-Green Lifesavers optical phenomenon. To observe this, look in a mirror in a dark room and chew the candy with your mouth open!

Slender false-wintergreen (*G. ovatifolia*)

Hairy false-wintergreen
(*G. hispidula*)

Hairy false-wintergreen (*G. hispidula*)

123

EDIBILITY: highly edible

FRUIT: Fruit mealy to pulpy, fleshy, berry-like capsules with a mild, wintergreen flavour.

SEASON: Flowers May to June. Fruit ripens late August into September.

DESCRIPTION: Delicate, creeping evergreen shrublets. Leaves leathery, small. Flowers white to greenish or pinkish.

WARNING: *Oil-of-wintergreen (most concentrated in the berries and young leaves) contains methyl salicylate, a drug that has caused accidental poisonings. It should never be taken internally, except in very small amounts. Avoid applying the oil when you are hot, because dangerous amounts could be absorbed through the open pores of your skin. It is known to cause skin reactions and severe (anaphylactic) allergic reactions. People who are allergic to aspirin should not use false-wintergreen or its relatives.*

Alpine false-wintergreen (*G. humifusa*) has 1–2 cm-long glossy leaves, rounded to blunt at tips with pinkish- or greenish-white, 5-lobed, 3–4 mm-wide flowers and scarlet, pulpy, 5–6 mm-wide berries. Berries are drier and not as palatable as other *Gaultheria* species. Grows in moist to wet, subalpine to alpine meadows in southern BC. Also called: alpine wintergreen, creeping wintergreen.

Hairy false-wintergreen (*G. hispidula*) has tiny, stiff, flat-lying, brown hairs on its stems and lower leaf surfaces. Leaves very small (4–10 mm long). Flowers tiny (2 mm wide), 4-lobed. Berries white, small (generally less than 5 mm in diameter), on a very short stalk, persisting through fall. Grows in cold, wet bogs and coniferous forests in montane and subalpine zones. Associated with acidic soils and often grows in mosses under conifers or on rotting logs, along the edges of swamps or bogs. Also called: creeping snowberry.

Alpine false-wintergreen (*G. humifusa*)

Hairy false-wintergreen (*G. hispidula*)

Alpine false-wintergreen (*G. humifusa*)

Slender false-wintergreen (*G. ovatifolia*) is similar to alpine false-wintergreen, but has reddish, hairy (rather than hairless) calyxes and pointy-tipped 2–5 cm-long leaves. Grows in moist to wet forests, heaths and bogs in montane and subalpine sites in southern BC.

Slender false-wintergreen (*G. ovatifolia*)

Slender false-wintergreen (*G. ovatifolia*)

Common bearberry (*A. uva-ursi*)

Bearberries are rather mealy and tasteless, but they are often abundant and remain on branches all year, so they can provide an important survival food. Many Canadian First Nations traditionally used them for food. To reduce the dryness, bearberries were often cooked with salmon oil, bear fat or fish eggs, or they were added to soups or stews. Sometimes, boiled berries were preserved in oil and served whipped with snow during winter. Boiled bearberries, sweetened with syrup or sugar and served with cream, reportedly make a tasty dessert. They can also be used in jams, jellies, cobblers and pies, or dried, ground and cooked into a mush. Apparently, if the berries are fried in grease over a slow fire they eventually pop, rather like popcorn. Scalded mashed berries, soaked in water for an hour or so,

Hairy manzanita (*A. columbiana*)

Alpine bearberry (*A. alpina*)

EDIBILITY: edible, not palatable

FRUIT: Small 5–10 mm fruits, bright red to purplish black.

SEASON: Flowers May to July. Berries ripen August to September.

DESCRIPTION: Evergreen or deciduous shrubs with clusters of nodding, white or pinkish, urn-shaped flowers and juicy to mealy, berry-like drupes containing 5 small nutlets. The genus *Arctostaphylos* contains 2 main groups of species, the bearberries, which are low, trailing to tufted shrubs found most abundantly in arctic and alpine regions, and the manzanitas, which are erect or spreading, taller shrubs of western Canada.

Red bearberry (*A. alpina* var. *rubra*)

produce a spicy, cider-like drink, which can be sweetened and fermented to make wine.

Although fairly insipid, juicy alpine bearberries are probably among the most palatable fruits in the genus, but because they grow at high elevations and northern latitudes, they have been the least used. Hairy manzanita berries are very similar to the ones of common bearberry, and were historically an important food for many tribes in the plant's range. Hikers sometimes chew the berries and leaves to stimulate saliva flow and relieve thirst.

Red bearberry (*A. alpina* var. *rubra*)

Common bearberry (*A. uva-ursi*)

Alpine bearberry (*A. alpina*)

Common bearberry (*A. uva-ursi*)

Alpine bearberry (*A. alpina*) is a trailing, deciduous shrub, to 15 cm tall. Leaves thin, veiny, oval, 1–5 cm long, with hairy margins (at the base) that often turn red in autumn, the previous year's dead leaves usually evident. Flowers small (4–6 mm long), appearing from June to July and producing mealy fruits that are purplish black, 5–10 mm in diameter, by late summer. Grows in moderately well-drained, rocky, gravelly and sandy soils on tundra, slopes and ridges in northern parts of the province. Also called: black alpine bearberry, whortleberry • *Arctous alpina*. Note: the common name "whortleberry" is also a common name for a totally unrelated species, *Vaccinium myrtillus*.

Common bearberry (*A. uva-ursi*) is a trailing, evergreen shrub to 15 cm tall. Leaves leathery, evergreen, spoon-shaped, 1–3 cm long. Small (4–6 mm long) flowers appear from May to June and produce bright red, 5–10 mm diameter, mealy fruits by late summer. Grows in well-drained, often gravelly or sandy soils in open woods and rocky, exposed sites at all elevations throughout BC. Also called: kinnikin-nick • *Arctous rubra*.

Red bearberry (*A. alpina* var. *rubra*) is similar to alpine bearberry, but it has longer leaves (to 9 cm long) with hairless margins, the leaves of the previous year not persistent. Berries bright red. Grows in the same habitats as alpine bearberry and over the same range.

Hairy manzanita (*A. columbiana*) is a much taller shrub, 1–3 m tall. Leaves 2–5 cm long, oval, evergreen. Fruits red-coloured. Grows in open, coniferous forests and openings in extreme southwestern BC.

Hairy manzanita (*A. columbiana*)

Hairy manzanita (*A. columbiana*)

Red bearberry (*A. alpina* var. *rubra*)

Salal *Gaultheria shallon*

Also called: Oregon wintergreen

Salal (*G. shallon*)

These dark, juicy berries grow in many places on the Pacific coast and were traditionally the most plentiful and important fruit gathered by Aboriginal peoples. The fruit was eaten fresh, dried into cakes for winter, or used as an important trade item. The Kwakwaka'wakw ate the ripe berries, dipped in eulachon grease, at large feasts. For trading, selling or food for common people, ripe salal berries were often mixed with currants, elderberries or unripe salal berries. Cakes of pure salal berries were saved to be eaten by immediate families or by chiefs at feasts. The berries were also used to sweeten other foods,

and the Haida used salal berries to thicken salmon eggs.

In recent times, salal berries have been prepared as jams or preserves, and the ripe berries from prime bushes are hard to beat for flavour and juiciness. They can be used in all recipes that traditionally call for blueberries, for example in syrups, pancakes, muffins, cookies and fruitcake. To make a refreshing drink, crush the berries and cook them, adding an equal measure of water. Strain (use a fine cloth as salal seeds are quite small), add a bit of sugar and/or lemon juice to enhance the flavour if desired, and chill the resulting mixture.

The fruit, which quickly stains your fingers while picking, also makes a good purple dye. The glossy green leaves of salal are widely harvested for export and sold to florists worldwide for use in floral arrangements. If you are gathering these fruit with children, pick a single berry off its stem, and gently squeeze the stem end between your thumb and index finger. If your chosen berry is a ripe and juicy one, this pressure will cause the blossom end to open out into a flower-shaped fruit. It's a fun riddle to make a "flower" out of a berry—and it's tasty to eat!

EDIBILITY: highly edible

FRUIT: Fruits purplish black, berry-like, to 1 cm in diameter.

SEASON: Flowers March to June. Fruit ripens August to September.

DESCRIPTION: Erect to partially creeping, freely branching evergreen shrub, to 3 m tall and often forming thickets. Leaves broadly oval, dark green, glossy, 3–9 cm long, 1–6 cm wide and finely serrated, dark green on top and lighter beneath. Leaf stalks, flower stems, bracts and young branches reddish and hairy, sometimes sticky. Flowers white to pinkish, urn-shaped, 5-lobed, 7–9 mm long, in 5- to 15-flowered clusters. An abundant shrub over much of the Pacific coast, growing in dry to wet forests, bogs and openings to montane elevations in coastal BC and, much less commonly, the southern Interior. Often grows on rotting logs and stumps.

Black Crowberry *Empetrum nigrum*

Also called: moss berry, curlew berry

Black crowberry (*Empetrum nigrum*)

Next to cranberries and blueberries, crowberries are one of the most abundant edible wild fruits found in northern Canada and were a vital addition to the diet of northern First Nations. Because they are almost devoid of natural acids, they can taste a little bland and were often mixed with blueberries or lard or oil and in more modern times with sugar and lemon. Their taste does seem to vary greatly with habitat—the flavour of the berries has been described in a range from bland, to tasting like turpentine, and even most delicious. Their taste improves after freezing or cooking, however, and their sweet flavour peaks after a frost.

The fruit is high in vitamin C, about twice that of blueberries, and is also rich in antioxidant anthocyanins (the pigment that gives them their black colour). Their high water content was a blessing to hunters seeking to quench their thirst in the waterless high country. As the berries have a firm, impermeable skin and are not prone to becoming soggy, they are ideal for making muffins, pancakes, pies, jellies (with added pectin), preserves and the like. For a fine dessert, cook the berries with a little lemon juice and serve them with cream and sugar.

Crowberries are usually collected in autumn, but because they often persist on the plant over winter, they can be picked (snow depth permitting) through to spring if the wildlife doesn't get them first. They are small fruit, so it can take up to 1 hour to pick 2 cups of berries! Consuming too many berries alone may cause constipation, so these berries have historically been prescribed for diarrhea. The berries make a reasonable black dye.

EDIBILITY: highly edible

FRUIT: Black, shiny, berry-like drupes to 9 mm. Contains large, inedible seeds.

SEASON: Flowers May to August. Fruits ripen July to November.

DESCRIPTION: Evergreen dwarf or low shrub, 5–10 cm tall, prostrate and mat-forming, to 30 cm long. Leaves dark to yellow-green to wine-coloured, 2–6 mm long, alternate but growing so closely together as to appear whorled, needle-like. Flowers inconspicuous, 1–3, pink, in leaf axils, 3 petals and sepals, petals 3 mm long, with male and female flowers separate but on the same plant. Fruit a juicy, black, berry-like drupe, 3–6 mm in diameter, containing 2–9 seeds, sometimes overwintering. Grows prolifically in bogs, moist shady forests, low-lying headlands, dry, acidic, rocky or gravelly soil on slopes, ridges and seashores in tundra, muskeg and spruce forests at all elevations.

Elderberries   *Sambucus* spp.

Blue elderberry (*S. nigra* ssp. *caerulea*)

R aw elderberries are generally considered inedible and cooked berries edible (see Warning), but some tribes are said to have eaten large quantities fresh from the bush. Cooking or drying destroys the rank-smelling, toxic compounds. Most elderberries were consumed after steaming or boiling, or were dried for winter use. Sometimes clusters of fruit were spread on beds of pine needles in late autumn and covered with more needles and eventually with an insulating layer of snow. These caches were easily located in the winter months by the bluish-pink stain they left in the snow! Only small amounts were eaten at a time, though, just enough to get a taste. Sometimes elderberries were steamed with black hair lichen for flavouring. Today, they are used in jams, jellies, syrups, preserves, pies and wine. Because these fruits contain no pectin, they are often mixed with tart, pectin-rich fruits such as crab apples.

Elderberries are also used to make teas and to flavour some wines (e.g., Liebfraumilch) and liqueurs (e.g., Sambuca). A delicious, refreshing fizzy drink called elderflower pressé or cordial can be made from the flowers. Red elderberry juice was sometimes used to marinate salmon prior to baking. The flowers can be used to make tea or wine, and in some areas, flower clusters were popular dipped in batter and fried as fritters or stripped from their relatively bitter stalks and mixed into pancake batter.

Red elderberry
(*S. racemosa* ssp. *pubens*)

Red elderberry
(*S. racemosa* ssp. *pubens*)

Blue elderberry (*S. nigra* ssp. *caerulea*)

Elderberries are rich in vitamin A, vitamin C, calcium, potassium and iron. They have also been shown to contain antiviral compounds that could be useful in treating influenza. The berries can be used to produce a brilliant crimson or violet dye. Elderberry wine, elderberries soaked in buttermilk and elderflower water have all been used in cosmetic washes and skin creams. The scientific name for *Sambucus* derives from the Greek instrument *sambuke*, in reference to the hollow pithy stems of this plant, which have been used in many different cultures to make musical instruments.

EDIBILITY: edible, edible with caution (toxic)

FRUIT: Fruits juicy, berry-like drupes, 4–6 mm across, in dense, showy clusters.

SEASON: Blooms April to July. Fruits ripen July to September.

DESCRIPTION: Unpleasant-smelling, 1–3 m tall, deciduous shrubs with pithy, opposite branches often sprouting from the base. Leaves pinnately divided into 5–9 sharply toothed leaflets about 5–15 cm long. Flowers white, 4–6 mm wide, forming crowded, branched clusters.

Blue elderberry (*S. nigra* ssp. *caerulea*) has flat-topped flower clusters and dull blue fruits with a whitish bloom. Leaves leaflets, usually 9. Grows in gravelly, dry soils on streambanks, field edges and woodlands.

Blue elderberry (*S. nigra* ssp. *caerulea*)

Red elderberry (*S. racemosa* ssp. *pubens*)

Blue elderberry (*S. nigra* ssp. *caerulea*)

Red elderberry (*S. racemosa* ssp. *pubens*) has pyramid-shaped flower clusters and shiny fruits, and is considered the tastiest of the genus. We have 2 common varieties here in BC: **black elderberry** (var. *melanocarpa*), with purplish-black fruit, grows predominantly in the Interior of BC, although it does occasionally occur west of the Coast Mountain range, and **red elderberry** (var. *pubens*), a coastal species with red fruit. Both grow in open woods, forest edges and roadsides as well as montane and subalpine sites. Also called: *S. melanocarpa, S. pubens.*

Red elderberry (*S. racemosa* ssp. *pubens*)

Black elderberry
(*S. racemosa* var. *melanocarpa*)

Blue elderberry (*S. nigra* ssp. *caerulea*)

Black elderberry
(*S. racemosa* var. *melanocarpa*)

WARNING: *All parts of this plant, except for the fruit and flowers, are considered toxic. The stems, bark, leaves and roots contain poisonous cyanide-producing glycosides (especially when fresh), which cause nausea, vomiting and diarrhea, but the ripe fruits and flowers are edible. The seeds, however, contain toxins that are most concentrated in red-fruited species. Many sources classify red-fruited elderberries as poisonous and black- or blue-fruited species as edible.*

Bush Cranberries   *Viburnum* spp.

High bush cranberry (*V. edule*)

Raw bush cranberries are high in vitamin C and can be very sour and acidic (much like true cranberries), but many native peoples ate them, chewing the fruit, swallowing the juice and spitting out the tough skins and seeds. Berries were traditionally mixed with grease and stored in birchbark baskets for winter use, trade or a valuable gift. They were also eaten with bear grease, or in an early year they could be mixed with sweeter berries such as saskatoons. Some tribes ate the boiled berries mixed with oil and occasionally this mixture was whipped with fresh snow to make a frothy dessert.

Bush cranberries are an excellent winter-survival food, because they remain on the branches all winter and are high enough that the snow doesn't cover them. Berries are best picked in autumn, after they have been softened and sweetened by a frost. Some people compare their fragrance to that of dirty socks, but the flavour is good (perhaps a Stilton of the berry world?). The addition of lemon or orange peel to the fruit, however, is said to eliminate this odour.

High bush cranberry (*V. edule*)

Today, bush cranberries are usually boiled, strained (to remove the seeds and skins) and used in jams and jellies. While these preserves usually require additional pectin (especially after the berries have been frozen), there are reports that imperfectly ripe berries (not yet red) gel without added pectin. Bush cranberry juice can be used to make a refreshing cold drink or fermented to make wine, and the fresh or dried berries can be steeped in hot water to make tea. Unfortunately, their large stones and tough skins limit their use in muffins, pancakes and pies. The berries produce a lovely reddish-pink dye and the acidic juice can be used as a mordant (required to set dyes and make the colour permanent). American bush cranberry makes a wonderful garden ornamental that is drought tolerant and provides not only pretty and scented spring flowers, but a showy fall foliar display and important winter wildlife food—if the humans don't get there first!

American bush cranberry (*V. opulus*)

EDIBILITY: edible, edible with caution (toxic)

FRUIT: Juicy, strong-smelling, red to orange berry-like drupes 1–1.5 cm long, with a single flat stone.

SEASON: Flowers April to July. Fruits ripen September to October.

DESCRIPTION: Deciduous shrubs with opposite, 3-lobed leaves, 3–12 cm long. Flowers white, small, 5-petalled, forming flat-topped clusters rather like a lacecap hydrangea. Leaves turn a showy red colour in fall.

American bush cranberry (*V. opulus*)

American bush cranberry (note that this is the same common name sometimes attributed to *V. edule*) (*V. opulus*) is a large 1–4 m tall shrub with a wide-spreading habit. Leaves maple-like with three relatively deeply cut lobes. Flower clusters resemble a lacecap hydrangea with a showy outer ring of large (12–25 mm wide), white, sterile flowers surrounding a central growth of tiny petal-less blooms. Grows in moist soils, hedges, scrub areas, plains and woodlands. Also called: high bush cranberry • *V. trilobum* var. *americanum*.

High bush cranberry (*V. edule*) is a scraggly-looking shrub 0.5–3.5 m tall. Bark smooth, grey with a reddish tinge. Leaves opposite, 6–12 cm wide, sharply toothed and hairy underneath, often with three shallow lobes evident toward the leaf tip. Flowers small, relatively inconspicuous clusters growing beneath leaf pairs. Its distinctive, musty smell may announce its presence before it is actually seen. It grows in shady foothills, damp woods, streambank thickets, and some montane and subalpine sites. Also called: mooseberry, squashberry, low bush cranberry.

American bush cranberry (*V. opulus*)

High bush cranberry (*V. edule*)

High bush cranberry (*V. edule*)

WARNING: *Some sources classify raw bush cranberries as poisonous, while others report that they were commonly eaten raw by native peoples. A few berries may be harmless, but large quantities can cause vomiting and cramps especially if they are not fully ripe, so it is probably best to cook the fruit before eating. Despite the common name of "cranberry," these species are not botanically related to the sour red berries we traditionally enjoy with a Thanksgiving feast.*

American bush cranberry (*V. opulus*)

High bush cranberry (*V. edule*)

American bush cranberry (*V. opulus*)

Soapberry — *Shepherdia canadensis*

Also called: russet buffaloberry

Soapberry (*S. canadensis*)

Soapberries were an important fruit for the First Nations within the plant's range, either for home use or as a trade item. The berries were eaten fresh, or they were boiled, formed into cakes and dried for future use. Because their juice is rich in saponin, soapberries become foamy when beaten. The ripe fruit can be mixed about 4:1 with water and whipped like egg whites to make a foamy dessert called "Indian ice cream." The resulting foam is truly unexpected and remarkable, having a beautiful white to pale pink colour and a smooth, shiny consistency of the best whipped meringue! Traditionally, this dessert was beaten by hand or with a special stick with grass or strands of bark tied to one end, these tools eventually being replaced by eggbeaters and mixers. Like egg whites, soapberries will not foam in plastic or greasy containers. The incredibly thick foam is rather bitter, so it was usually sweetened with sugar or with other berries. Soapberries can also be added to stews or cooked to make syrup, jelly, jam or a sauce for savoury meats. Canned soapberry juice, mixed with sugar and water, makes a refreshing "lemonade." Although they are bitter, soapberries are often abundant and can be used in moderation as an emergency food (see Warning).

The fruit was collected by beating the branches over a canvas or hide and then rolling the berries down a wooden board into a container to separate leaves and other debris.

Soapberries are rich in vitamin C and iron. They have been taken to treat flu and indigestion and have been made into a medicinal tea for relieving constipation. Canned soapberry juice, mixed with sugar and water, was used to treat acne, boils, digestive problems and gallstones. Soapberry bark tea was a favourite solution for eye troubles. These berries can also be crushed or boiled to use as a liquid soap.

WARNING: *This species contains saponin, a bitter, soapy substance that can irritate the stomach and cause diarrhea, vomiting and cramps if consumed in large amounts.*

EDIBILITY: highly edible

FRUIT: Bright red oval berries with a fine silvery scale, juicy.

SEASON: Flowers April. Fruits ripen July to September.

DESCRIPTION: Deciduous shrub, open-formed, to 2 m tall. Young twigs covered in a brown or rusty scale. Older twigs and branches brownish red with orange flecks, sometimes fissured. Leaves somewhat thick, elliptic, smooth-edged, tip rounded, opposite, top green with short silvery scales, rusty underneath when young. Flowers yellowish to greenish, male and female flowers on separate plants, single or forming small clusters. Fruit grows on very short stalks on female plants, at leaf axils. Grows in open woods, mixed forests, and on streambanks. Prefers moist habitat but will tolerate some drought.

Indian Ice Cream

Makes approximately 6 cups

Even with sugar this treat will have a slightly bitter taste, but many people quickly grow to like it.

1 cup soapberries • 1 cup water
4 Tbsp granulated white sugar

Put berries and water into a wide-topped ceramic or glass mixing bowl. *Do not use a plastic bowl or utensils, and make sure that nothing is greasy, or the berries will not whip properly.* Whip the mixture with an electric eggbeater or hand whisk until it reaches the consistency of beaten egg whites. Gradually add the sugar to the pink foam, but not too fast or the foam will "sink." Serve immediately.

Silverberry   *Elaeagnus commutata*

Also called: wolf willow

Silverberry (*E. commutata*)

The berries of this species are very dry and astringent, but some northern tribes gathered them for food. Most groups considered the mealy berries famine food and did not ingest them regularly. When eaten, they were consumed raw or cooked in soup. They were also cooked with blood, mixed with lard and eaten raw, fried in moose fat or frozen. Despite not being very palatable raw, they reportedly make good jams and jellies and the berries are apparently much sweeter after exposure to freezing temperatures. Some tribes used the nutlets inside the berries as decorative beads. The fruits were boiled to remove the flesh, and while the seeds were still soft, a hole was made through each. They were then threaded, dried, oiled and polished.

The flowers can be detected from metres away by their sweet, heavy perfume. Some people enjoy this fragrance, but others find it overwhelming and nauseating. If green wolf willow wood is burned in a fire, it gives off a strong smell of human excrement! Some practical jokers enjoy sneaking branches into the fire and watching the reactions of fellow campers.

EDIBILITY: not palatable

FRUIT: Fruits silvery, mealy, about 1 cm long, drupe-like, with a single large nutlet.

SEASON: Blooms June to July. Fruits ripen in late August to September.

DESCRIPTION: Thicket-forming, rhizomatous shrub with 2–6 cm-long, alternate, lance-shaped, deciduous, silvery leaves covered in dense, tiny, star-shaped hairs (appearing silvery). Flowers strongly sweet-scented, yellow inside and silvery outside, 6–16 mm long, borne in twos or threes from leaf axils. Grows on well-drained, often calcareous slopes, gravel bars and forest edges at low to montane elevations.

145

One-flowered clintonia (*C. uniflora*)

The berry of this species, though very pretty and unusual to look at, is dry, tasteless and mildly toxic so is not recommended for eating. The Nuxalk of Bella Coola called this berry "wolf's berry" because it was considered inedible to humans and edible only to wolves. The Thompson people of the Interior traditionally made a dye from the fruit. This plant, which spreads through a network of underground rhizomes, matures to form a pretty combination of light green leaves with delicate flowers and metallic-blue berries, and is well worth growing for its ornamental value, particularly when planted with other low-growing natives like bunchberry (white flowers and bright red berry clusters) and trillium (striking white to pink blooms). Birds such as blue grouse relish the berries.

EDIBILITY: edible with caution (toxic)

FRUIT: Shiny, dark blue, metallic-looking "bead" to 9 mm growing atop a single stem.

SEASON: Flowers May to June. Fruits ripen July to August.

DESCRIPTION: Spreading perennial herb arising from rhizomes, to 50 cm tall and often forming colonies. Leaves basal, 2–3 on each plant, 2–5 cm wide, shiny, narrowed at both ends, to 30 cm long. Flowers single, very occasionally double. Flowering stem slender, 14–40 cm tall, erect, usually hairy at the top. Berries single, rarely double, round to oblong, coloured bright metallic blue, 6–12 mm thick. Found in shaded, moist to mesic forests and open woods.

Twisted-stalks *Streptopus* spp.

Rosy twisted-stalk (*S. lanceolatus*)

These perennials are called "twisted-stalks" because of the kinks (sometimes right-angled, sometimes just a sharp curve) present in the main stem or flower stalks. Most native peoples regarded twisted-stalks as poisonous and used the plant mainly for medicine, but some tribes ate young plants and/or the bright-coloured berries, either raw or cooked in soups and stews. The berries are juicy and moderately sweet tasting but are mildly toxic so should only be eaten in small quantities. Indeed, eating more than a few reportedly causes diarrhea and it is best to consider these berries inedible.

Twisted-stalks were highly regarded for their general restorative qualities and were taken as a tonic or to treat general sickness. The whole plant was taken by some First Nations people to treat coughs, loss of appetite, stomach-aches, spitting up blood, kidney trouble and gonorrhea. The blossoms were ingested to induce sweating. The plant was sometimes tied to, and used to scent, the body, clothes or hair.

First Nations' names for the berries included owl berries, witch berries, black bear berries and frog berries; the berries were also believed to be eaten by snakes, deer and wolves.

Twisted-stalks differ from the closely related fairybells in that the flowers attach to the stem in the leaf axils instead of to the branch tip. The characteristic branched stem (sometimes zigzagging) is what separates clasping twisted-stalk from the other twisted-stalks.

Clasping twisted-stalk
(*S. amplexifolius*)

Rosy twisted-stalk
(*S. lanceolatus*)

Small twisted-stalk
(*S. streptopoides*)

EDIBILITY: edible with caution (toxic)

FRUIT: Berries hanging, red-orange or yellowish, egg-shaped and somewhat translucent; seeds small, whitish, somewhat visible.

SEASON: Blooms late June to early August. Fruits ripen August to September.

DESCRIPTION: Slender, herbaceous perennial, from thick, short rhizomes, 0.4–1 m tall or more. Leaves smooth-edged, elliptical- or oval-shaped, alternate, markedly parallel-veined. Flowers small, white, bell-shaped, 8–12 mm long, with 6 petals that flare backward, each hanging on the lower side of each stalk, 1 per leaf. Found in moist shaded forests, clearings, meadows, disturbed sites and on streambanks at low to subalpine elevations throughout the province.

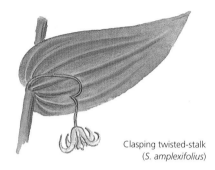

Clasping twisted-stalk
(*S. amplexifolius*)

Clasping twisted-stalk (*S. amplexifolius*)

Clasping twisted-stalk (*S. amplexifolius*)

Rosy twisted-stalk (*S. lanceolatus*)

Clasping twisted-stalk (*S. amplexifolius*) grows 0.5–1 m tall and has branched, smooth stems, sometimes bent at nodes, giving it a zigzag appearance. Leaves clasping at base. Flowers greenish white. Berries bright red.

Rosy twisted-stalk (*S. lanceolatus*) grows to 30 cm tall, with stems usually unbranched, curved (not zigzagged), leaves not clasping; rose-purple or pink flowers with white tips; red berries. Also called: *S. roseus*.

Small twisted-stalk (*S. streptopoides*) grows 10–20 cm tall. Leaves oval-shaped to oblong-lance shaped, smooth, 3–6 cm long and 1.5–2 cm wide. Flowers wine-coloured with a yellowish-green tip, single hanging from leaf axils, 1–5 per stem. Fruit orange to red, globose, 5–6 mm in size. Inhabits damp, dense coniferous forests.

WARNING: *Young twisted-stalk plants closely resemble green false-hellebore, which is **extremely** poisonous. Collecting the young shoots of twisted-stalk for consumption is not recommended unless you are absolutely sure of plant identification!*

Rosy twisted-stalk (*S. lanceolatus*)

Clasping twisted-stalk (*S. amplexifolius*)

Small twisted-stalk
(*S. streptopoides*)

151

Lilies-of-the-valley *Maianthemum* spp.

False lily-of-the-valley (*M. dilatatum*)

Berries of both species mentioned here are considered edible but are bitter-tasting and not very palatable.

Many groups in BC and neighbouring areas ate false lily-of-the-valley, but it was rarely highly regarded. The berries were usually only eaten casually by children or by hunters and berry pickers while out on trips. Some groups, such as the Haida, used the berries to a great extent. They ate these berries fresh or picked them when unripe and stored them until they were red and soft. Green berries were sometimes cooked in tall cedar boxes lowered into boiling water for a few minutes, and then the cooked fruit was mixed with other fruits before being sun-dried into cakes. Sometimes berries were scalded and eaten with animal or fish grease or stored this way. The fruit of false lily-of-the-valley was used as a medicine for tuberculosis.

Caution is advised when eating berries of wild lily-of-the-valley as eating too many can cause severe diarrhea. In a Haida myth, a feast for supernatural beings included wild lily-of-the-valley berries. Wild lily-of-the-valley was given the name "frog berry" by the Kwakwaka´wakw because it was said that these amphibians ate the berries.

The genus name is derived from the Latin word for "May," referring to the flowering time of these plants. The fruit of *Maianthemum* species is a true berry in botanical terms.

EDIBILITY: not palatable

FRUIT: Berries pea-sized, at first hard and green, soft and red when ripe.

SEASON: Flowers May to June. Fruits ripen July to September.

DESCRIPTION: Herbaceous creeping perennial herbs arising from rhizomes and usually forming large colonies. Leaves heart-shaped, alternating, usually 2 or 3, with prominent parallel veins. Flowers small, white, with 4 petals, 4–6 mm wide, borne in distinct terminal clusters, blooming early spring. Fruit borne at the top of stems, in clusters. Found in moist woods and clearings.

False lily-of-the-valley (*M. dilatatum*)

False lily-of-the-valley (*M. dilatatum*) is a larger plant, stems 10–25 cm tall, normally with two leaves near the top and one near the base. Leaves stalked. Berries hard and green when unripe, turning mottled brown then soft and red when ripe. Occurs in swampy areas and shady, moist woods in coastal BC with a limited range in the east-central part of the province. Also called: two-leafed Solomon's-seal.

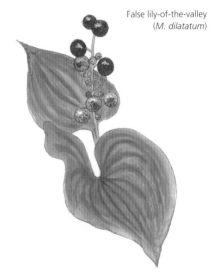

False lily-of-the-valley
(*M. dilatatum*)

False lily-of-the-valley (*M. dilatatum*)

Wild lily-of-the-valley (*M. canadense*)

False lily-of-the-valley (*M. dilatatum*)

Wild lily-of-the-valley (*M. canadense*) is a smaller plant, less than 25 cm tall; leaves stalkless. Berries cream coloured with red speckles, then pinkish with red flecks, ripening to a solid red. Found in eastern BC in moist forests. Also called: Canada mayflower.

WILD GARDENING: *This species grows into a delightful, dense carpet of delicate heart-shaped leaves and makes an excellent low-maintenance understorey planting for the woodland garden. The flowers are pretty in spring and the berries provide a showy late summer and fall display that attracts wildlife such as grouse that like to eat them.*

Wild lily-of-the-valley (*M. canadense*)

Wild lily-of-the-valley
(*M. canadense*)

155

False Solomon's-seal (*M. racemosum*)

Various indigenous peoples across Canada ate the ripe berries, young greens and fleshy rhizomes of this species. Some BC First Nations believed the berries to be the food of snakes and avoided them. In cases where berries were eaten, it was usually casually (hunters, berry pickers, children). The Gitxsan, however, picked them ripe in August, preserved them in eulachon grease and stored them in boxes in a cool place for winter eating. They were said to be reserved as food for chiefs. The Carrier called the fruits sugarberry and considered them sweet and good to eat. Berries of star-flowered false Solomon's-seal are said to be high in vitamin C.

EDIBILITY: not palatable

FRUIT: See individual species descriptions.

SEASON: Flowers May to June. Berries ripen August to October.

Star-flowered false Solomon's-seal
(*M. stellatum*)

False Solomon's-seal
(*M. racemosum*)

Star-flowered false Solomon's-seal
(*M. stellatum*)

Star-flowered false Solomon's-seal
(*M. stellatum*)

DESCRIPTION: Tall, herbaceous perennials growing from thick, whitish, branching rhizomes, often found in dense clusters. Leaves smooth-edged, broad, elliptical, alternate along stems in 2 rows, 5–15 cm long, distinctly parallel-veined, often clasping. Flowers small, cream-coloured, 6-parted, in dense, terminal clusters. Berries small and densely clustered, initially green and mottled or striped, ripening to bright red. Grows in rich woods, thickets and moist clearings.

False Solomon's-seal (*M. racemosum*) grows in clumps to 1.2 m tall from a fleshy, stout rootstock. Stems unbranching, arching. Flowers in clusters of 50–70, berries a tight cluster of many seedy berries. Berries at first green with copper spots, ripening to red, often with purple spots. Also called: feathery false lily-of-the-valley • *Smilacina racemosa*.

Star-flowered false Solomon's-seal (*M. stellatum*) grows to 50 cm tall, flowers in clusters of 5–6, berries at first green with blue-purple stripes. Differentiated from *M. racemosum* by being smaller, fewer flowers and leaves, and a lot fewer berries (2–8) that are larger, and green with red stripes when unripe. Also called: starry false lily-of-the-valley • *Smilacina stellata*.

Fairybells *Prosartes* spp.

Rough-fruited fairybells (*P. trachycarpa*)

The berries of this species were not widely eaten by First Nations and many BC tribes considered them poisonous. However, the Thompson and Shuswap ate the berries raw; rough-fruited fairybells were called false raspberries in the Shuswap language. Some reports describe the fruit of Hooker's fairybells as somewhat sweet tasting and juicy, but growing so sparsely as to not warrant the effort to gather them. The fruit of the rough-fruited fairybell has been described as distinctly apricot flavoured. Reports vary greatly as to whether the berries of fairybells are edible or not, so caution is advised.

Fairybells were associated with ghosts or snakes by some First Nations. Common First Nations names for these fruit include "snake berries," and "grizzly bear's favourite food." Rodents and grouse are known to feed on the berries. The leaves of this species are "drip tips," a form that ingeniously channels rainwater to the base of the plant.

Hooker's fairybells (*P. hookeri*)

Hooker's fairybells (*P. hookeri*) has smooth berries with 4–6 seeds. Also called: Oregon fairybells • *Disporum hookeri.*

Rough-fruited fairybells (*P. trachy-carpa*) has stamens that hang well below the petals of the flowers, and conspicuously rough-skinned, velvety-surfaced orange to red berries with 6–12 seeds. Also called: *Disporum trachycarpum.*

Hooker's fairybells (*P. hookeri*)

EDIBILITY: edible

FRUIT: Berries egg-shaped, orange or yellow to bright red.

SEASON: Flowers April to July. Berries ripen July.

DESCRIPTION: Perennial herbs, 30–60 cm tall, with few branches, from thick-spreading rhizomes. Leaves alternate, broadly oval, 3–9 cm long, pointed at tips, rounded to a heart-shaped base, fringed with short, spreading hairs, prominently parallel-veined. Flowers creamy to greenish white, narrowly bell-shaped, 1–2 cm long, drooping, 1–3 at branch tips. Grows in moist woods, forests and thickets as well as subalpine meadows.

Rough-fruited fairybells (*P. trachycarpa*)

Rough-fruited fairybells (*P. trachycarpa*)

Strawberries *Fragaria* spp.

Woodland strawberry (*F. vesca*)

These delicious little berries pack significantly more flavour than a typical large, domestic strawberry. Wild strawberries are small compared to modern cultivars and are probably best enjoyed as a nibble along the trail, but they can also be collected for use in desserts and beverages. A handful of bruised berries or leaves, steeped in hot water, makes a delicious tea, served either hot or cold.

Strawberries were popular berries with all BC First Nations, but their juiciness can make them difficult to dry and preserve. Today strawberries are prepared by freezing, canning or making jam, but traditionally, they were sun-dried. The berries were mashed and spread over grass or mats to dry in cakes, which were later eaten dry or rehydrated, either alone or mixed with other foods as a sweetener. Anyone who has had the extreme pleasure of savouring dried wild strawberries knows that this is a treat well worth the time to prepare! Strawberry flowers, leaves and stems were sometimes mixed with roots in cooking pits as a flavouring.

Strawberries contain many quickly assimilated minerals (e.g., sodium, calcium, potassium, iron, sulphur and silicon), as well as citric and malic acids, and they were traditionally prescribed to enrich the bloodstream. Strawberry leaf tea, accompanied by fresh strawberries, was a recommended remedy for gout, rheumatism, inflamed mucous membranes and liver, kidney and gallbladder problems. Strawberries are a good source of ellagic acid, a chemical that is believed to prevent cancer. To remove tartar and whiten discoloured teeth, strawberry juice can be held in the mouth for a few minutes and then rinsed off with warm water. This treatment is reported to be most effective with a pinch of baking soda in the water. Large amounts of fruit in the diet also appear to slow dental plaque formation. Strawberry juice, rubbed into the skin and later rinsed

Woodland strawberry (*F. vesca*)

Virginia strawberry
(*F. virginiana*)

Virginia strawberry (*F. virginiana*)

off with warm water, has been used to soothe and heal sunburn.

Many people will be surprised to learn that the strawberry is technically *not* a fruit! What we think of as the "berry" is actually a swollen receptacle (this is the base of the flower, which you would normally expect to see inside a fruit). The true "fruits" are the tiny dark seeds (seed-like achenes) which one can easily find either embedded in, or perched on, the soft flesh of the strawberry.

The Virginia strawberry is the only species of strawberry found on Haida Gwaii, and is increasingly rare owing to over-browsing by introduced deer. Virginia strawberry and beach strawberry are the original parents of 90 percent of our modern cultivated strawberry varieties.

EDIBILITY: highly edible

FRUIT: Berries are red when ripe, resembling miniature cultivated strawberries.

SEASON: Flowers bloom May to August. Fruits ripen starting in June, and flowers continue to bloom throughout this season, so plants often have ripe fruit and flowers on them at the same time.

DESCRIPTION: Low creeping perennials with long, slender runners (stolons). Leaves green, often turning red in the fall, 5–10 cm across, with 3 sharply toothed leaflets. Flowers white, 5-petalled, 1.5–2 cm across, usually several per stem, forming small, loose clusters.

Virginia strawberry (*F. virginiana*)

Beach strawberry (*F. chiloensis*)

Virginia strawberry (*F. virginiana*)

Beach strawberry (*F. chiloensis*) has green, thick, leathery leaflets, with the end tooth shorter than its adjacent teeth. Veins are netted and leaf surface is wrinkled above and hairy below. Plants produce hairy runners. Found on dunes, rock crevices, and sea bluffs and beaches only along coastal BC. Also called: coastal strawberry.

Virginia strawberry (*F. virginiana*) has bluish green leaflets, with the end tooth narrower and shorter than its adjacent teeth. Common throughout BC in dry to moist open woodlands

and clearings, often in disturbed areas on well-drained sites in prairie to subalpine zones, but most often found in the Interior. Also called: wild strawberry, common strawberry.

Woodland strawberry (*F. vesca*) has yellowish-green leaflets, with the end tooth projecting beyond its adjacent teeth. Leaflets are thick, hairy, strongly veined and scalloped. Found in dry to moist open woods and meadows and on streambanks commonly through southern BC and more rarely in the northern portion of the province.

Virginia strawberry (*F. virginiana*)

Woodland strawberry (*F. vesca*)

Wild Berry Muffins

Makes 12 muffins

This batter can also be baked in a loaf form.

5 Tbsp vegetable oil • 2 eggs, lightly beaten
1½ cups mixed wild berries (strawberries, thimbleberries, blueberries, huckleberries, etc.)
1 tsp salt • 1¾ cups whole wheat flour
¾ cup brown sugar • 2¼ tsp baking powder

Preheat oven to 400° F. Mix wet ingredients together in a bowl. Sift dry ingredients together in another bowl. Make a shallow well in the centre of the dry ingredients and slowly add the wet mixture. Mix well and pour into greased or lined muffin tins. Bake for 10 to 15 minutes, or until a knife inserted into a muffin comes out clean.

Bunchberry *Cornus canadensis*

Also called: Canada dogwood, dwarf dogwood • *C. unalaschensis*

Bunchberry (*C. canadensis*)

The bright scarlet-orange fruits of this woodland plant look like they should be very poisonous but are actually quite edible. However, opinions of their flavour range from insipid to a pulpy, sweet, flavourful fruit similar in taste to the highly regarded salal berry. Coastal First Nations such as the Bella Coola and Sechelt gathered bunchberries to eat fresh, either on their own or with eulachon grease, and more recently enjoyed them with granulated sugar. While the Haida considered them not as tasty as other berries, they still sometimes gathered them to steam and preserve with water and grease for winter consumption.

The berries are abundant where the plant grows and it is easy to gather a quantity with minimal effort. They can be eaten raw as a trail nibble and are also said to be good cooked in puddings. However, each drupe contains quite a hard seed so be wary of mature dental work. Bunchberries (often mixed with other fruits) can be used whole to make sauces and preserves or cooked and strained to make beautiful scarlet-coloured syrups and jellies.

Bunchberry is reported to have anti-inflammatory, fever-reducing and pain-killing properties (rather like mild aspirin), but without the stomach irritation and potential allergic effects of salicylates. The plant has a history of being used to treat headaches, fevers, diarrhea, dysentery and inflammation of the stomach or large intestine. The berries were eaten and/or applied in poultices to reduce the potency of poisons. They were also chewed and the resulting pulp applied topically to soothe and treat burns. Bunchberries have historically been steeped in hot water to make a medicinal tea for treating paralysis, or boiled with tannin-rich plants (such as common bearberry or commercial tea) to make a wash for relieving bee stings and poison-ivy rash. Native peoples used tea made with the entire plant to treat aches and pains, lung and kidney problems, coughs, fevers and fits.

EDIBILITY: edible

FRUIT: Bright orange-red berry-like drupes, 6–9 mm wide, growing in dense clusters at the stem tips, nestled into a whorl of leaves (hence the common name "bunchberry"). The drupe has a yellowish pulp and single seed.

SEASON: Flowers from May to August. Fruit ripens July to August.

DESCRIPTION: Perennial, rhizomatous herb, 5–20 cm tall. Leaves wintergreen-like, 2–8 cm long, growing opposite each other in groupings of 2–6, spaced so tightly that they have a whorled appearance. Flowers tiny, in a dense clump at the centre of 4 white to purple-tinged, petal-like bracts (exactly like miniature flowers of the dogwood tree), forming single, flower-like clusters about 3 cm across. Grows in cool, moist woods and damp clearings at low to subalpine elevations, commonly found on rotting stumps and logs.

AMAZING: *This plant spreads its relatively heavy pollen grains through an interesting "explosive pollination mechanism." When the pollen is ripe and ready to be released, an antenna-like trigger lets go in the flower, rapidly springing the four pollen-laden anthers violently upward together in a snapping motion, thereby catapulting the pollen grains far up into the air for dispersal.*

English Holly *Ilex aquifolium*

English holly (*I. aquifolium*)

Introduced from Eurasia, this species is a common garden and municipal ornamental throughout coastal BC. Its bright red berries are a traditional symbol of the Christmas holidays; indeed, BC has a thriving holly export industry that sends the sprigs of leaves and berries to the rest of Canada and parts of the U.S. where the winter temperature is too extreme for this species to grow. Considered a threat to BC's endangered Garry oak ecosystem, holly has few natural predators here. It is well-protected from browsing by its spiky leaves, and its seeds are spread far and wide by the birds, such as robins, that love eating its berries.

EDIBILITY: poisonous

FRUIT: Clustered round berries (drupes), to 9 mm, containing 4 pits, shiny red (sometimes orange or yellow), staying on the tree throughout winter and often well into spring.

SEASON: Flowers May. Fruits ripen September.

DESCRIPTION: Evergreen tree growing to 15 m. Bark smooth, grey. Leaves to 12 cm long, 2–6 cm broad, glossy and leathery, variable in shape, alternate, single. Often, but not always with large, spiny teeth, pointing alternately upward and downward. Flowers scented, clustered, bell-shaped, 4-petalled, white or pale green to 6 mm long. Male and female trees separate, with only the females bearing fruit. Widespread throughout mild coastal BC, preferring moist forests, clearings and edges but also inhabiting more marginal habitat such as dry forest.

WARNING: *Holly berries contain ilicin, a compound irritating to intestines and the stomach as well as harmful to the heart and nervous system. All holly berries are poisonous, and ingesting them can cause nausea and diarrhea, and in quantity (more than 20) can trigger violent vomiting or even death.*

Cascara *Rhamnus purshiana*

Cascara (*R. purshiana*)

Some sources report that native peoples ate the purple berries of these small trees or shrubs. However, this was most likely in modest amounts given their strong purgative action; some sources consider these fruit poisonous. The berries have been described as bland to bitter in taste, and should not generally be considered edible.

This species has been used as a laxative for at least hundreds of years (and probably longer given First Nations historic use) and is highly considered since its action is fairly gentle and is non-habit forming. Indigenous peoples collected the bark in spring and summer. It was then dried and stored for later use. The dried bark was traditionally used to make medicinal teas, but today it is usually administered as a liquid extract or elixir, or in tablet form. Each year, 0.5–4 million kg of bark is collected commercially, mainly from wild trees in BC, Washington, Oregon and California.

EDIBILITY: edible with caution (toxic), poisonous

FRUIT: Berry-like drupes 6–9 mm long, often ripening unevenly in bunches, turning from green to yellow to a purple or black colour.

SEASON: Flowers June to July. Ripens August to September.

DESCRIPTION: Erect or spreading deciduous shrub or small tree, to 10 m tall. Leaves alternate, glossy, oval to elliptic, prominently veined with smooth or finely toothed margins, 5–12 cm long and turning a pretty yellow colour in the fall. Bark is smooth, grey and bitter tasting. Flowers are 5-petalled, inconspicuous, greenish yellow, all male or all female in 1 stalked cluster, forming flat-topped clusters of up to 25 flowers in leaf axils. Grows on streamsides and in open to closed forests at low to montane elevations in southern BC.

WARNING: *Some sources consider these fruit poisonous. This genus contains large amounts of anthraquinones, which are responsible for its emetic properties. Ingesting fresh bark and berries can have very severe effects, but curing the bark for at least 1 year or using a heat treatment reduces the harshness.*

Pacific Yew *Taxus brevifolia*

Also called: western yew, mountain mahogany

Pacific yew (*T. brevifolia*)

The bark of Pacific yew is an original source of the important anti-cancer drug taxol. After a long period of development by the National Cancer Institute and pharmaceutical partners, this drug was approved for use in treating a variety of cancers and is particularly successful in treating breast and ovarian cancers that historically had extremely low survival rates.

The slow-growing western yew, however, became quickly depleted in the wild by unregulated over-harvesting. There are ongoing concerns regarding its natural regeneration since it was also removed for many years as a "weedy" species in second growth timber stands, and it requires both a male and female tree growing in relative proximity to each other to reproduce.

Taxane derivatives are now in great part obtained from managed harvest of the more common Canada yew (*Taxus canadensis*) in eastern Canada, from which the drug is prepared by extraction and semi-synthesis. Some native peoples used yew bark for treating illness (indeed, this is how modern researchers first knew to research this plant).

EDIBILITY: edible with **extreme** caution (toxic), poisonous

FRUIT: Berry-like arils, 4–5 mm across, with a cup of orange to red fleshy tissue around the single bony seed. The showy, berry-like fruit of this species, with its sweet taste but slimy texture, has historically been considered edible. However, the hard seeds found within the fleshy cup are extremely poisonous, so this fruit is not recommended for consumption (see Warning).

SEASON: Flowers in June. Berries ripen to an orange or deep red in August to October.

DESCRIPTION: Small, generally scraggly looking evergreen shrub or tree, with a straight trunk, to 15 m tall (rarely to 25 m). Branches drooping, bark reddish brown, scaly and flaking. Often grows together in small thickets. Needles soft, flattened, 3.5 cm long, arranged alternately in two rows, glossy green above and paler green underneath with two whitish bands of stomata and a sharp tip. Male and female are separate trees, the male pollen-bearing cones inconspicuous. The tiny green flowers of the female tree eventually produce scarlet red arils. Unusually, this conifer has no pitch. Grows in moist, shady sites such as streambanks and under mature coniferous forest, at low to montane elevations in BC.

WARNING: *The needles, bark and seeds contain extremely poisonous, heart-depressing alkaloids called taxanes. Drinking yew tea or eating as few as 50 leaves can cause death. Many birds eat the berries (which take 2 years to mature) and the branches are said to be a preferred winter browse for moose, but many horses, cattle, sheep, goats, pigs and deer have been poisoned from eating yew shrubs, especially when the branches were previously cut.*

171

Pacific poison-oak (*T. diversilobum*)

Poison-ivy and poison-oak plants contain an oily resin (urushiol) that causes a nasty skin reaction in most people, especially on sensitive skin and mucous membranes. Since the allergic contact dermatitis appears with some delay after exposure many people do not realize that they have come into contact with these plants until it is too late. Sensitization can also lead to more severe reaction after repeated exposure.

Urushiol is not volatile and therefore is not transmitted through the air, but it can be carried to unsuspecting victims on pets, clothing and tools and even on smoke particles from burning poison-ivy or poison-oak plants. The resin can persist on pets and clothing for months and is also ejected in fine droplets into the air when the plants are pulled.

Washing with a strong soap can prevent a reaction if it is done shortly after contact since this process can remove the resin. Washing also prevents transfer of the resin to other parts of the body or to other people. Be sure to use cold water as warm water can help the resin to penetrate into your skin where it is extremely difficult to remove. The liquid that oozes from poison-ivy or poison-oak blisters on affected skin does not contain the allergen. Ointments and even household ammonia can be used to relieve the itching of mild cases, but people with severe reactions might need to consult a doctor.

Western poison-ivy (*T. rydbergii*)

Western poison-ivy (*T. rydbergii*)

EDIBILITY: poisonous

FRUIT: Fruits whitish to brown, berry-like drupes 4–5 mm wide.

SEASON: Flowers May to June. Fruits ripen July to August.

DESCRIPTION: Trailing to erect deciduous shrubs, forming colonies. Leaves bright glossy green, resinous, compound, divided into 3 oval leaflets, scarlet in autumn. Flowers cream coloured, 5-petalled, 1–3 mm across, forming clusters. Separate male and female plants.

Pacific poison-oak (*T. diversilobum*)

Pacific poison-oak (*T. diversilobum*) grows to 2 m tall and has round-tipped leaflets that are usually lobed (hence the reference to "oak") and shorter than poison-ivy's (to 7 cm long vs. to 15 cm or more in poison-ivy). Fruits whitish drupes. It grows on drier, rocky slopes at low elevations on southeastern Vancouver Island and the nearby Gulf Islands.

Western poison-ivy (*T. rydbergii*), to 2 m tall, spreads mainly by stolons and so forms distinct patches. Its leaflets are entire (neither toothed nor lobed). Western poison-ivy is a species of drier rocky slopes, most commonly in the BC Interior.

Devil's Club Oplopanax horridus

Also called: *Echinopanax horridum*

Devil's club (*O. horridus*)

Throughout its range, devil's club (which is botanically related to the ginseng family) is considered to be one of the most powerful and important of all medicinal plants. First Nations people of coastal BC considered the berries of this plant to be inedible, perhaps partly because they are held aloft above a remarkable fortress of irritating spiny leaves and stems and because even the berries have spikes! Although devil's club tea is recommended today for binge-eaters who are trying to lose weight, some tribes used it to improve appetite and to help people gain weight. In fact, it was said that a patient could gain too much weight if it was used for too long. Some tribes used a strong decoction of the plant to induce vomiting in purifying rituals preceding important events such as hunting or war expeditions. This decoction was also applied to wounds to combat staphylococcus infections, and ashes from burned stems were sometimes mixed with grease to make salves to heal swellings and weeping sores. Like many members of the ginseng family, devil's club contains glycosides that are said to reduce metabolic stress and thus improve one's sense of wellbeing. The roots and bark of this plant contain the majority of active compounds, and have traditionally been used in the treatment of arthritis, diabetes, rheumatism, digestive troubles, gonorrhea and ulcers. The root tea has been reported to stimulate the respiratory tract and to help bring up phlegm when treating colds, bronchitis and pneumonia. It has been used to treat diabetes because it helps regulate blood sugar levels and reduce the craving for sugar. Indeed, devil's club extracts have successfully lowered blood sugar levels in laboratory animals.

Possibly because of its diabolical spines, devil's club was considered a highly powerful plant that could protect one from evil influences of many kinds. Devil's club sticks were used as protective charms, and charcoal from the burned plant was used to make protective face paint for dancers and others who were ritually vulnerable to evil influences. The Haida rubbed the bright red berries into hair to combat dandruff and lice and to add shine. The Bella Coola people believed that these berries were a favourite food of grizzly bear, and thus called the plant "grizzly-bear berries."

WARNING: *Devil's club spines are brittle and break off easily, embedding in the skin and causing infection. Some people have an allergic reaction to the scratches from this plant. Wilted leaves can be toxic so only fresh or completely dried leaves should be used to make a medicinal tea, but even then the tea should be taken under the guidance of a registered herbalist and in moderation because extended use can irritate the stomach and bowels.*

EDIBILITY: poisonous, and not recommended

FRUIT: Fruits bright red, berry-like drupes, slightly flattened, sometimes spiny, 5–8 mm long, in showy pyramidal terminal clusters.

SEASON: Flowers May to July. Fruits ripen July to September.

DESCRIPTION: Strong smelling, deciduous shrub, 1–3 m tall, with spiny, erect or sprawling stems. Leaves broadly maple-like, 10–40 cm wide, with prickly ribs and long, bristly stalks. Spines on leaves and stems grow up to 9 mm long. Flowers greenish white, 5–6 mm long, 5-petalled, forming 10–25 cm-long, pyramid-shaped clusters of bright red berries. Grows in moist, shady coastal areas, foothills and montane sites. Easy to find when you lose your footing on a coastal trail as it is invariably the plant that you grab onto to stop your fall.

Honeysuckles *Lonicera* spp.

Black twinberry (*L. involucrata*)

Although some honeysuckle berries are nauseously bitter, others are mildly pleasant. The shiny black fruit of black twinberry, which grows as a shrub rather than a vine, were not eaten by First Nations and were considered poisonous by many. However, wildlife such as ravens, bears and crows eat them regularly and a few reports claim that the berries are not harmful in small quantities. They are very bitter tasting, though, so are not very palatable.

Honeysuckle has been used by a number of First Nations for a diverse range of ailments. The leaves of orange honey-suckle were prepared and used as a contraceptive and for any problems of the womb. An infusion of the twigs was ingested, for colds and sore throats and in small amounts for epilepsy. Chewed leaves were applied as a poultice to heal bruises. Pregnant women ingested a tea made from the berries and bark to expel worms. Black twinberry leaves were chewed and applied externally to itchy skin, burns, inflammation, boils and gonorrheal sores. The fruit was also crushed and rubbed in the hair to treat dandruff. In spite of the myriad traditional medical uses of honeysuckle, however, they are rarely used today in modern herbalism.

First Nations used the stems of twining honeysuckles as building materials and to make fibres for mats, baskets, bags, blankets and toys. Twining honeysuckle plants were also employed in a number of love potions and charms. Crushed black twinberry fruit was used to make a dye for basketry materials.

EDIBILITY: edible with caution (toxic), poisonous

FRUIT: See individual species descriptions.

SEASON: Flowers May to July. Fruit ripens August to September.

DESCRIPTION: Twining, woody shrub or vine, with opposite leaves, to 10 cm long. Flowers sweet-scented, bell-shaped, 2–4 cm long. Fruit an orange, black or red berry about 1 cm across containing several seeds.

Orange honeysuckle (*L. ciliosa*) is a widely branching vine growing to 6 m long. Flowers produce a sweet nectar from May to July, grow in clusters at the stem tip and are orange-yellow to red. Leaves are long-hairy on the margins. Fruit are a translucent orange-red, grow to 1 cm each and contain several seeds. Grows in mesic to dry forests and thickets at low to montane elevations

Orange honeysuckle (*L. ciliosa*)

in extreme southern BC. Also called: western trumpet honeysuckle.

Twining honeysuckle (*L. dioica*) is a twining vine growing to 5 m long. Flowers are clustered at the stem tips, yellow to orange in colour (sometimes dark reddish with age), and produce a sweet, edible nectar from May to July. Leaves are not hairy on the margins. Fruit are dark red, 8–12 mm across. Grows in dry woods, thickets and rocky slopes in eastern and southern BC. Also called: glaucous-leaved honeysuckle, limber honeysuckle, red honeysuckle, smooth-leaved honeysuckle.

Black twinberry (*L. involucrata*) is a deciduous, erect shrub, growing to 5 m tall with 4-angled twigs that are greenish when young, greyish with shredding bark when older. Leaves oval, to 16 cm long and 8 cm wide, sharp-pointed at the tip. Flowers (which do not contain an edible nectar) are bell-shaped, yellow, 1–2 cm long, in pairs rather than clusters, surrounded by fused bracts. Grows in moist or wet soils in forests, clearings, riverbanks, swamps and thickets across the majority of the province. Also called: twinflower honeysuckle.

Twining honeysuckle (*L. dioica*)

Common Snowberry *Symphoricarpos albus*

Common snowberry (*S. albus*)

Although some sources report that these berries are edible, though not very good, snowberries are toxic and in large quantities can be mildly poisonous. Most tribes considered snowberries poisonous and did not eat them but the Haida reportedly consumed them. Some believed snowberries were the ghosts of saskatoons, part of the spirit world and not to be eaten by the living.

The spongy white berries are fun to squish and pop—rather like bubble wrap! The unusual white berries persist on the plant through the winter, providing a showy and decorative display that in mild winters can last well into spring. The berries make a wonderful addition to winter holiday wreaths, garlands and other festive decorations. This is a very drought-tolerant and decorative species that will thrive on steep slopes and other areas that may otherwise be difficult to landscape. The leaves and flowers, albeit small, are pretty and the white berries provide winter forage for birds and small mammals while giving a showy winter display.

Common snowberry is considered a management concern in threatened Garry oak habitat (Canada's most endangered ecosystem, with less than 1 percent of the original extent remaining) since it quickly forms dense thickets that smother other plant species. Removing snowberry is difficult and labour intensive, and pulling out the dense root masses causes a great deal of soil disturbance and damage to surrounding plants. Because of its drought tolerance, tenacious roots and thick growth habit, however, it is an ideal planting to stabilize slopes and provide vegetation in difficult to plant areas.

EDIBILITY: edible with caution (toxic), poisonous

FRUIT: White, waxy, spongy, berry-like drupes, 6–10 mm long, singly or in clusters on stem tips.

SEASON: Flowers May to August. Berries ripen and whiten August and September.

DESCRIPTION: Erect, deciduous shrub usually 50–75 cm tall. Leaves pale green, opposite, elliptic to oval, 2–4 cm long. Stems flexible and strong, grey in colour, with bark becoming shredded on more mature specimens. Flowers pretty pink to white, broadly funnel-shaped, 4–7 mm long, borne in small clusters at the stem tips. Spreads rapidly through a tough, dense, underground root system (rhizomes) and quickly forms an impenetrable thicket if left alone. Grows throughout BC on rocky banks, hedgerows, forest edges and roadsides.

WARNING: *The branches, leaves and roots of this plant are poisonous, containing the alkaloid chelidonine, which can cause vomiting, diarrhea, depression and sedation.*

Red Baneberry *Actaea rubra*

Also called: snake berry • *A. arguta, A. eburnea*

Red baneberry (*A. rubra*)

Baneberry is related to the commercial phytomedicine black cohosh, and some indigenous peoples used baneberry root tea in a similar way to treat menstrual and postpartum problems, as well as colds, coughs, rheumatism and syphilis. Herbalists have used baneberry roots as a strong antispasmodic, anti-inflammatory, vasodilator and sedative, usually for treating menstrual cramps and menopausal discomforts.

Baneberry is a striking-looking plant with its attractive foliage and delicate stems of puffy, white flowers in spring, followed by showy spikes of red or white berries in the fall. Planted with ferns, hostas and other shade-loving species, it makes for a decorative addition to the shade garden.

EDIBILITY: poisonous

FRUIT: Fruits are very showy, several-seeded, glossy, red or white berries 6–8 mm long, growing singly on a long stalk.

SEASON: Flowers May to July. Ripens July to August.

DESCRIPTION: Branched, leafy, generally solitary perennial herb, 30 cm–1 m tall, from a woody stem-base and fibrous roots. Stems long, wiry. Leaves are coarsely toothed, alternate, few and large, divided 2–3 times in threes, crowded at base of stem and sparser near the top. Flowers white, with 5–10 slender, 2–3 mm-long petals, forming long-stalked, rounded clusters with many flowers. Inhabits deciduous forests, mixed coniferous forests, subalpine meadows, moist woodlands, streambanks and swamps at low to montane elevations.

WARNING: *All parts of baneberry are poisonous, but the roots and berries are most toxic. Indeed, the common name "baneberry" derives from the Anglo-Saxon* bana, *which means "murderous." Eating as few as 2–6 berries can cause severe cramps and burning in the stomach, vomiting, bloody diarrhea, increased pulse, headaches and/or dizziness. Severe poisoning results in convulsions, paralysis of the respiratory system and cardiac arrest. No deaths have been reported in North America, probably because the berries are extremely bitter and unpleasant to eat.*

Glossary

accessory fruit: a fruit that develops from the thickened calyx of the flower rather than from the ovary (e.g., soapberry).

achene: a small, dry fruit that doesn't split open; often seed-like in appearance; distinguished from a nutlet by its relatively thin wall.

alkaloid: any of a group of bitter-tasting, usually mildly alkaline plant chemicals. Many alkaloids affect the nervous system.

alternate: situated singly at each node or joint (e.g., as leaves on a stem) or regularly between other organs (e.g., as stamens alternate with petals).

anaphylaxis: a hypersensitivity reaction to the ingestion or injection of a substance (a protein or drug) resulting from prior contact with a substance. Anaphylaxis can progress rapidly and be life-threatening.

annual: a plant that completes its life cycle in one growing season.

anthers: the pollen-producing sacs of stamens.

anthraquinone: an organic compound found in some plants that has a laxative effect when ingested. It is also used commercially as a dye and pigment, and also in the pulp and paper industry.

aril: a specialized cover attached to a mature seed.

armed: a plant furnished with defensive bristles or thorns.

axil: the position between a side organ (e.g., a leaf) and the part to which it is attached (e.g., a stem).

berry: a fleshy, simple fruit that contains one or more ovule-bearing structures (carpels) that each contains one or more seeds; the outside covering (endocarp) of a berry is generally soft, moist and fleshy (e.g., blueberry).

biennial: a plant that lives for two years, usually producing flowers and seed in the second year.

bitters: alcoholic drinks consumed with a meal which contain bitter herbs to aid in the process of digestion (e.g., Swedish bitters or Angostura bitters).

Vascular Plant Parts

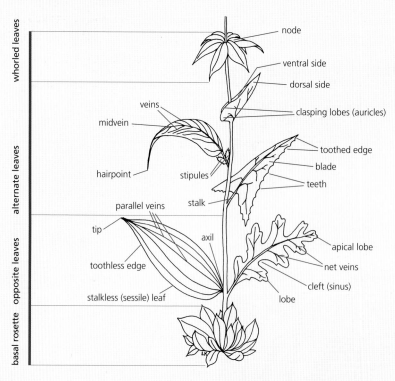

bog: a peat-covered wetland characterized by *Sphagnum* mosses, heath shrubs and sometimes trees.

bract: a specialized leaf with a flower (or sometimes a flower cluster) arising from its axil.

calcareous: a type of soil with a high calcium content.

calyx: the outer (lowermost) circle of floral parts; composed of separate or fused lobes called sepals; usually green and leaf-like.

carpel: a fertile leaf bearing the undeveloped seed(s); one or more carpels join together to form a pistil.

cathartic: a substance that purges the bowels.

compound leaf: a leaf composed of two or more leaflets.

compound drupe: a collection of tiny fruit that forms within the same flower from individual ovaries; this type of fruit is often crunchy and seedy (e.g., boysenberries).

cone: a fruit that is made up of scales (sporophylls) that are arranged in a spiral or overlapping pattern around a central core, and in which the seeds develop between the scales (e.g., juniper).

corolla: the second circle of floral parts, composed of separate or fused lobes called petals; usually conspicuous in size and colour, but sometimes small or absent.

cultivar: a plant or animal originating in cultivation (e.g., loganberry or Golden Delicious apple).

deciduous: having structures (leaves, petals, seeds, etc.) that are shed at maturity and in autumn.

drupe: a fruit with an outer fleshy part covered by a thin skin and surrounding a hard or bony stone that encloses a single seed (e.g., a plum).

drupelet: a tiny drupe; part of an aggregate fruit such as a raspberry.

emetic: induces vomiting.

endocarp: the inner layer of the pericarp.

eulachon grease: grease from the eulachon (*Thaleichthys pacificus*), a small species of fish in the smelt family that lives most of its life in the Pacific Ocean but comes inland to fresh water for spawning.

fruit: a ripened ovary, together with any other structures that ripen with it as a unit.

glabrous: without hair, smooth.

glandular: associated with a gland (e.g., glandular hair).

glaucous: a frosted appearance due to a whitish powdery or waxy coating.

globose: shaped like a sphere.

glycoside: a two-parted molecule composed of a sugar and an aglycone, usually becoming poisonous when digested and the sugar is separated from its poisonous aglycone.

habitat: where a plant or animal is normally found; the characteristic environmental conditions in which a species is normally found.

haw: the fruit of a hawthorn, usually with a fleshy outer layer enclosing many dry seeds.

heath: a member of the heath family (Ericaceae).

herbaceous: a plant or plant part lacking lignified (woody) tissues.

hip: a fruit composed of a collection of bony seeds (achenes), each of which comes from a single pistil, covered by a fleshy receptacle that is contracted at the mouth (e.g., rose hip).

hybrid: a cross between two species.

hybridize: breeding together different species or varieties of plants or animals; the resulting hybrid often has characteristics of both parents.

inflorescence: flower cluster.

involucre: a set of bracts closely associated with one another, encircling and immediately below a flower cluster.

lanceolate: a long leaf that is widest at the middle and pointed at the tip.

lenticel: a slightly raised pore on root, trunk or branch bark.

mesic: habitat with intermediate moisture levels—not too dry or too moist.

montane: mountainous habitat, below the timberline.

multiple fruit: ripens from a number of separate flowers that grow closely together, each with its own pistil (e.g., mulberry, fig).

node: the place where a leaf or branch is attached.

nutlet: a small, hard, dry, one-seeded fruit or part of a fruit; does not split open.

opposite: situated across from each other at the same node (not alternate or whorled); or situated directly in front of another organ (e.g., stamens opposite petals).

ovary: the part of the pistil that contains the ovules.

ovules: the organs that develop into seeds after fertilization.

palmate: divided into three or more lobes or leaflets diverging from a common point, like fingers on a hand.

peduncle: a flower or fruit stem.

pemmican: a mixture of finely pounded dried meat, fat and sometimes dried fruit.

perennial: a plant that lives for three or more years, usually flowering and fruiting for several years.

pericarp: the part of a fruit that derives from the ovary wall; generally consists of three layers: (from inside to outside) endocarp, mesocarp, exocarp.

petal: a unit of the corolla; usually brightly coloured to attract insects.

phytomedicine: the use of plants as medicine.

pinnate: with branches, lobes, leaflets or veins arranged on both sides of a central stalk or vein; feather-like.

pistil: the female part of the flower, composed of the stigma, style and ovary.

pitch: sticky tree sap, for example from a pine tree.

pome: a fleshy fruit with a core (e.g., an apple) comprised of an enlarged hypanthium around a compound ovary.

prostrate: growing flat along the ground.

purgative: causing watery evacuation of the bowels.

raceme: an unbranched cluster of stalked flowers on a common, elongated central stalk, blooming from the bottom up.

receptacle: an expanded stalk tip at the centre of a flower, bearing the floral organs or the small, crowded flowers of a head.

recurved: curved under (usually referring to leaf margins).

rhizome: an underground, often lengthened stem; distinguished from the root by the presence of nodes and buds or scale-like leaves.

saponin: any of a group of glycosides with steroid-like structure; found in many plants; causes diarrhea and vomiting when taken internally but commercially used in detergents.

sepal: one segment of the calyx; usually green and leaf-like.

spore: a reproductive body composed of one or several cells that is capable of asexual reproduction (doesn't require fertilization).

sporophyll: a spore-bearing leaf; a scale of a conifer cone.

spp.: abbreviation of "species" (plural).

spur: a hollow appendage on a petal or sepal, usually functioning as a nectary.

spur-shoot: a slow-growing, much-reduced shoot (e.g., on a larch or ginko tree).

stolon: a slender, prostrate, spreading branch, rooting and often developing new shoots and/or plants at its nodes or at the tip.

style: the part of the pistil connecting the stigma to the ovary; often elongated and stalk-like.

subalpine: just below the treeline, but above the foothills.

sucker: a shoot not originating from a seed, but from a rhizome or root.

tepal: a sepal or petal, when these structures are not easily distinguished.

throat: the opening into a corolla tube or calyx tube.

toxic: a substance that can cause damage, illness or death.

tundra: a habitat in which the subsoil remains frozen year-round characterized by low growth and lacking in trees.

unarmed: without prickles or thorns.

variety: a naturally occurring variant of a species; below the level of subspecies in biological classification.

Section of a regular flower with numerous carpels

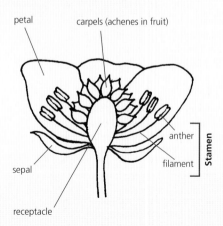

petal

carpels (achenes in fruit)

anther

filament

Stamen

sepal

receptacle

References

Bennet, Jennifer, Ed. 1991. *Berries: a Harrowsmith Gardener's Guide.* Camden House Publishing, Willowdale, Ontario.

Black, Marmelade. 1977. *It's the Berries.* Raintree Press, Inc & Hancock House Publishers Ltd., Saanichton, British Columbia.

Davis, A., B. Wilson & B. Compton. 1995. *Salmon Blossoms in the New Year.* Nanakila Press, Kitamaat, British Columbia.

Derig, Betty B., & Margaret C. Fuller. 2001. *Wild Berries of the West.* Mountain Press Publishing Company, Missoula, Montana.

Domico, Terry. 1979. *Wild Harvest: Edible Plants of the Pacific Northwest.* Hancock House Publishers Ltd., Saanichton, British Columbia.

Harrington, H. D. 1967. *Western Edible Wild Plants.* The University of New Mexico Press, Albuquerque, New Mexico.

Hutchens, Alma R. 1973. *Indian Herbalogy of North America.* Shambala Publications Inc., Boston & London.

Kershaw, Linda. 2000. *Edible and Medicinal Plants of the Rockies.* Lone Pine Publishing, Edmonton, Alberta.

Kuhnlein, Harriet V. & Nancy J. Turner. 1991. *Traditional Plant Foods of Canadian Indigenous Peoples: Nutrition, Botany and Use.* Gordon and Breach Science Publishers, Philadelphia, Pennsylvania.

MacKinnon, A., L. Kershaw, J. T. Arnason, P. Owen, A. Karst & F. Hamersley Chambers. 2009. *Edible and Medicinal Plants of Canada.* Lone Pine Publishing, Edmonton, Alberta.

MacKinnon A., J. Pojar & R. Coupe, Eds. 1992, 1998. *Plants of Northern British Columbia: Expanded Second Edition.* BC Ministry of Forests, Vancouver, British Columbia & Lone Pine Publishing, Edmonton, Alberta.

Moore, Michael. 1993. *Medicinal Plants of the Pacific West.* Red Crane Books, Inc., Santa Fe, New Mexico.

Nabhan, Gary Paul, Ed. 2008. *Renewing America's Food Traditions.* Chelsea Green Publishing Company, White River Junction, Vermont.

Parish, R., R. Coupe & D. Lloyd, Eds. 1996. *Plants of Southern Interior British Columbia and the Inland Northwest.* BC Ministry of Forests, Vancouver, British Columbia & Lone Pine Publishing, Edmonton, Alberta.

Schofield, Janice J. 1989. *Discovering Wild Plants (Alaska, Western Canada, The Northwest).* Alaska Northwest Books, Anchorage, Alaska & Seattle, Washington.

Stark, Raymond. 1981. *Guide to Indian Herbs.* Hancock House Publishers Ltd., North Vancouver, British Columbia.

Szczawinski, A. F. & G. A. Hardy. 1971. *Guide to Common Edible Plants of British Columbia.* British Columbia Provincial Museum, Victoria, British Columbia.

Tilford, Gregory L. 1997. *Edible and Medicinal Plants of the West.* Mountain Press Publishing Company, Missoula, Montana.

Turner, Nancy J. 1997. *Food Plants of Interior First Peoples.* UBC Press, Vancouver, British Columbia.

Turner, Nancy J & Adam F. Szczawinski. 1988. *Edible Wild Fruits and Nuts of Canada.* National Museum of Natural Sciences, Markham, Ontario.

Underhill, J. E. 1974. *Wild Berries of the Pacific Northwest.* Hancock House Publishers Ltd., Saanichton, British Columbia.

Internet Sources

Canadian Biodiversity: http://canadianbiodiversity.mcgill.ca/english/species/plants/index.htm

E-FLORA BC, Electronic Atlas of the Plants of British Columbia: http://www.geog.ubc.ca/biodiversity/eflora/

Evergreen Native Plant Database: http://nativeplants.evergreen.ca/

Flora of North America, from the Flora of North America Association: http://www.fna.org/FNA

Natureserve: http://natureserve.org/explorer/

United States Department of Agriculture Plants Database: Natural Resources Conservation Service: http://plants.usda.gov/

Index to Common and Scientific Names

Entries in **boldface** type refer to the primary species accounts.

Photo Credits

Photos: BC Parks Service 168; Lee Beavington 13, 15, 16, 17, 19ab, 22b, 23, 24, 26, 28, 29ab, 31abc, 32, 33ab, 34, 36, 37ac, 40, 42, 44, 46, 47a, 48abc, 51, 52, 53, 54, 56, 57, 58, 59ab, 64, 65, 66ab, 69a, 71, 74, 75, 77abc, 78, 79, 80a, 81, 82, 84b, 86c, 94, 95, 96, 97ab, 98, 100, 101ab, 102, 103ab, 105ab, 106, 107ab, 109, 110c, 111, 112, 113abc, 116a, 117, 118, 120ab, 121, 128, 130, 131ab, 134, 135ab, 136ab, 137a, 138, 139a, 140, 141b, 146, 147ab, 148, 149a, 150a, 151, 152, 154a, 156, 157a, 159a, 161b, 162a, 164, 165, 166, 167ab, 170, 171, 174, 175, 176, 177b, 179, 180, 181ab, 182, 183ab; Frank Boas 92b; Neil Jennings 61, 80b, 144, 145a; Krista Kagume 63, 157b; Linda Kershaw 18, 37b, 39, 45, 55ab, 69b, 86a, 127b, 129, 133, 139b, 149b, 150b, 154b, 158, 159b, 161a, 163, 177c; Brian Klinkenberg 125b; Ron Long 145b; Tim Matheson 30; Jim Pojar 86b, 127a, 149c, 177a; Virginia Skilton 14, 22a, 62, 88, 92a, 125a, 169, 173, 178; Robert D. Turner and Nancy J. Turner 10, 11, 25, 27, 35, 38, 41, 47b, 49, 50, 60, 68, 70, 72, 73, 76, 83, 84a, 87, 90, 91ab, 93ab, 104, 110ab, 116b, 119, 122, 123, 126, 132, 137b, 141ac, 142, 143, 155, 160, 162b; Bri Weldon 172.

Illustrations: All illustrations are by Frank Burman and Ian Sheldon, except: Linda Kershaw, 12, 13, 14, 184.

About the Author

FIONA HAMERSLEY CHAMBERS was born in Vancouver and spent much of her childhood around coastal BC. Most of her formative years were spent with her family at the old Dididaht village site of Clo-oose (in what is now Pacific Rim National Park Reserve in BC), in the UK countryside, and in the Coast Salish community on Kuper Island near Chemainus, BC. It is to this early experience in nature and with First Nations communities that she attributes her life-long interest in ethnobotany. She holds an undergraduate degree from the University of Victoria in French and Environmental Studies (1994), a Masters of Science in Environmental Change and Management from Oxford University (1998), and a Masters in Environmental Design from the University of Calgary (1999). Speaking English, French and Spanish, she has travelled extensively throughout Europe, Australia, New Zealand, Mexico and Central America and has a strong interest in learning about traditional plant uses wherever she goes. Fiona has taught Environmental Studies at the University of Victoria since 1999, and ethnobotany at Pacific Rim College since 2009. She currently divides her time between teaching, running a small organic farm and food plant nursery (www.metchosinfarm.ca), writing books and academic papers, working as a ship's naturalist (www.mapleleafadventures.com), and raising two energetic boys who also love plants, animals and bugs. She enjoys outdoor pursuits like hiking, camping and playing with her children, and is especially fond of foraging for wild foods to cook with her family back at home. She also plays the guitar and sings.